Machines – Code – People

Christian Abegg; Peter Gfader

Machines – Code – People

50 things Zühlke engineers are passionate about

Bibliographical Information of the Deutsche Nationalbibliothek
This publication is listed in the Deutsche Nationalbibliographie of the Deutsche Nationalbibliothek; detailed bibliographical information can be accessed under http://dnb.d-nb.de

© 2019 Christian Abegg, Peter Gfader
Printing, Production and Layout:
BoD – Books on Demand, Norderstedt, Germany

ISBN: 978-3-7481-4118-1

Table of contents

Foreword (Nicolas Durville) 9
Preface 11

Part I: Me   **13**

Coding on purpose (Adrian Herzog) 15
Fast tracking into new projects – take notes! (Sven Bayer) 20
If you like it then you shouldn't put some code in it (Jonathan Rothwell) 25
Never forget (Igor Spasić) 28
Pragmatic documentation (Ognjen Blagojević) 30
Rules are dangerous (Rolf Bruderer) 34
Start using a time management technique today (Christian Abegg) 38
Why you need to go visual (Gabriel Duss) 42
"You are not a software engineer" – What am I then? (Stefan Djelekar) 47

Part II: People   **51**

"As long as you live under my roof, you'll do as I say" – If the project manager leads differently than I would (Sabrina Lange) 53
Business readiness – is there readiness for agile development in business? (Ina Paschen) 56
CYA: Cover your ass (Romano Roth) 59
Digitalization and its impact on customer interaction (Beat Bischof) 62
Discipline flow (Stephan Janisch) 68
Don't teach kids programming (Igor Spasić) 74
From enthusiasm to commercial success (Bojan Jelaca) 76

Know-how transfer – just explaining once is not enough (Christoph Zuber). 79
Lean startup: taming the uncertainty (Eric Fehse, Sven Bünte, Stefan Reichert). 83
Meeting with users is essential for creating great products (Markus Flückiger). 87
Some inconvenient truths about the digitalization of your business (Moritz Gomm). 96
Team fit (Marko Simić). 103
The evangelist and the chameleon (Franziska Meyer). 106
The evolution of support and operations team setups (Tijana Krstajic, Guido Angenendt). 110
The house of the six wise men (Michael Richter) 116
Time to say goodbye (Sabrina Lange) 119
Transitioning systems engineering into the lean-agile world (Rolf P. Maisch). 124
We are all engineers but work quite differently: software engineers, electronics engineers, mechanics engineers (Thomas Weber). 130
What's wrong with: "I don't write any tests, since I am not a tester"? (Peter Gfader) 136
When machine learning meets software engineering (Wolfgang Giersche). 140
Why every project should have gardeners (Sabrina Lange) 143
Why you should create a paper prototype – and how to test it with your users (Eric Fehse, Manuel Jung). 147
You DiD what? (Marko Ivanović) 151
Your team needs a tech lead, not a lead techie (Daniel Mölle). 156

Part III: Machines, Code   163

Application first – a bottom-up architecture approach (Markus Rehrs). 165

Architectural programming (Stephan Janisch, Christian Eder, Alexander Derenbach). 169
Architectural programming in the development workflow (Stephan Janisch, Christian Eder, Alexander Derenbach). 175
CI and CD done right (Florian Besser) 178
Clean code best practices (Milan Milanović) 184
Codify your developer VMs! (Torben Knerr) 187
Containerisation and why to use it (Florian Besser). 191
Do something about that slow SQL query (Ognjen Blagojević). 195
Frontend is not your enemy (Janko Sokolović) 198
How to deal with flaky system tests (Adrian Herzog). 203
Making your tests run fast (Simon Lehmann) 208
Optimization and realization (Igor Spasić). .212
Rules for building systems (Vassilis Rizopoulos). 214
Successful agile system development with continuous system integration (Erik Steiner). 220
The best technology is not always the best choice (Carsten Kind) .223
Watch your state (Raphael M. Reischuk) 227
You always have time for a proper root cause analysis (Matthias Meid) 232

Foreword

I can still remember my early days as a software developer after my studies. It was during the wild years in the middle of the "new economy" and I had to get up to speed with new projects and acquire the necessary technical knowledge within a very short time. I had to gain a lot of experience for myself and I learned a great amount on the job; unfortunately, sometimes the hard way.

Later, as a project manager, I learned a great deal about how to collaborate with people and, above all, that, alongside technological knowledge, cooperation, leadership, working techniques, communication and relationship maintenance are very important. In my experience, projects often fail not because of the technology, but because of unclear or incorrect requirements, a lack of willingness to cooperate, sheer complexity or because the organization is simply overstrained.

The collection "Machines, Code, People" is a wonderful summary of what I have often experienced myself and what is really important in everyday project work. Code and machines together build a solution. Knowing these technologies is crucial. And to bring about real innovation, it is equally important to have the right people on board, to build up a great team and to really understand the needs of the customer.

This impressive collection was created as a joint effort within the Zühlke Group across all areas and it makes me extremely proud that so many authors have contributed so actively and with so much passion.

I hope it will inspire you and give you many insights that will benefit you in your daily work.

Nicolas Durville, CEO Zühlke Switzerland

Preface

Electronic version

This book is also available as eBook:

HTML: https://zuehlke.github.io/machines-code-people/
EPUB: https://zuehlke.github.io/machines-code-people/machines-code-people.epub
MOBI: https://zuehlke.github.io/machines-code-people/machines-code-people.mobi
PDF: https://zuehlke.github.io/machines-code-people/machines-code-people.pdf

Further reading and additional resources

Some articles refer to further reading and additional resources. As most of them are only available online, the printed version of this book contains QR codes referring to the HTML version of the article where the links to these external resources can be found.

PART I:
ME

Coding on purpose

When carrying out a project, it is essential to know its ultimate goal. And usually we start our projects exactly there, with some sort of *mission statement*, defining the *goal* of our endeavour. But it is all too easy to lose sight of this goal in the daily struggle of solving specific issues along the way. That's why many methodologies in software development try to help us by keeping the project aligned with its *purpose*.

- In **RUP**, one tries to establish traceability from lower level items to higher level goals.
- **Use cases** represent a goal at some level of abstraction. A use case diagram shows how lower level use cases support higher level use cases.
- **User stories** are often formulated according to a template that includes a *goal* and a *benefit* (e.g. "as <role> I want to <goal> so that <benefit>").
- With **user story maps** we sort *user stories* by how important they are in achieving a goal.
- An **impact map** traces down the high level goals to stakeholders, impacts and finally deliverables.
- **Design thinking** has a strong focus on *identifying and understanding customer needs*.
- The idea of the **minimal viable product** makes the point that we should only build as much as is necessary in order to achieve the desired goal.

If you've read up to here and have not thought of Simon Sinek yet, this can only be because you don't know him yet. He's currently the loudest voice when it comes to telling everybody to start with why. Look up his name if you have time for a YouTube evening.

Unfortunately, there are many areas that make no reference at all to the goal. Some examples:

- **Code** (and most other models of software, like UML diagrams), which is mostly imperative in its nature, just declares *how* something is achieved but does not state *why*. Although the purpose was hopefully clear to the person writing the code, it is likely to be unknown to the poor coder maintaining the code years later.
- I've seen too many **bad user stories** that either don't state a goal and benefit or if they do, it's just a wild guess by somebody who wanted to satisfy the template structure. And worse: higher level items like *epics* fail to mention the *purpose* even more often.
- **BDD (Behaviour Driven Design)** typically uses the *Given, When, Then* pattern, which only describes what the software does, but not why.
- Many consulting companies **get told what to build** instead of *what problem to solve* or *what outcome to achieve*.
- **Bad managers / leaders** might tell you *what* to build instead of *what outcome to strive for*.

When the solution becomes the problem

Sometimes you are immersed in a tricky problem. While some people might just give up, your passion for problem solving results in you pondering over the puzzle for hours and days, maybe even weeks. While working on the solution for your customer's problem, the solution suddenly turned into the problem. Or as Paul Watzlawick says:

> By searching for solutions we restrict ourselves with constraints that don't exist in the original problem.

When you're stuck in such a situation it's worth taking one or two steps back and asking: *what was the goal we wanted to achieve?* And also: *why did we want to achieve this goal in the first place?* This helps you increase the scope for possible solutions and hopefully allows you to discover that the problem you're trying to solve does not even need solving but instead you find a simpler, less problematic solution to the initial *why*. If you ask *why* five times, this approach even has a name: The five whys technique.

Measure impact, not output

Think about how you track the achievement of the desired outcome in your project. In the end, it's the job of your *Product Owner* or *Product Manager*, but if you think they could use some inspiration, they would probably be grateful for some suggestions. As an example, John Cutler proposes to add a column to your task board called *Achieved desired outcome*. And Gregor Ilg says:

> Don't celebrate when you have launched a product. Celebrate once you've learned from it.

Instead of measuring how many story points we can deliver per sprint, we should apply our maths skills much more to measuring and tracking whether we have achieved the desired outcomes in production. Do we achieve the desired user adoption? Has the desired percentage of business shifted to the new platform? Do we save the desired amount of manual work using the new system? And what do our end users tell us?

Empathise with the people that matter

Empathy and good collaboration with end users is certainly important for building a good user experience. But listening to the users is not enough.

You might hear a lot of different goals from different people in the project. Especially if there's a lack of official goals, people might fill them up with their own hidden or not so hidden agenda. Try to figure out who is incurring expense for the project and try to understand what outcomes they hope to achieve with this investment.

Lead through purpose

When leading a team, you should put a strong focus on communicating the *why* of the things that you or your team are asked to do, and judging your success by whether the *why* is fulfilled and not whether you built exactly *what* was asked for or whether you did it *how* it was expected that you should do it.

Coding tips

- If you get told *what* to do it's usually good to ask *what for*, as this gives you more flexibility in finding an economic solution for the desired outcome.
- When coding, be aware that code does not document the *why* particularly well. So be nice to your fellow coders and leave meaningful comments explaining the *why* or pointing to the relevant documentation.
- When asking somebody for help, tell them what problem you are

trying to solve, not just what problem you have with your current solution approach (see "The XY Problem").

Wrap up

If all of this was a bit much, here's one simple tip: no matter how minimalistic you want to keep your user stories, make sure they contain at least the desired impact.

By Adrian Herzog

Links:

Fast tracking into new projects – take notes!

Everybody in the software industry will come to a point when one has to enter a new project. Then, we need to familiarise ourselves with a different project environment and depend on other team members. Our decisions have to be thought out carefully until we become familiar with the project. Clearly, it is desirable to keep the transition-time as short as possible. A way of achieving this is by taking notes in a structured way. When we join a new project, this enables us to become an efficient team member as fast as possible.

In the past decades, more and more software projects have shifted to agile methodologies like Scrum. Because of that, the software industry requires employees to become more self-organised. Team members with this skill require less management and increase the project velocity on that way.

In this article, I show you how taking and organising notes, handwritten and digital, can help you to incorporate faster in new projects.

Ways to organise notes

Storing notes electronically on your laptop is the standard way to go. There are many different alternatives like Microsoft OneNote, Evernote, or your project's wiki. However, having a laptop on hand is not always an option. Sometimes you need to take notes by hand. There are different solutions, like a tablet with a pen or a paper notebook. In my optinion, the best combination is between a tablet or paper notebook and a laptop. For organising notes, I recommend using the outlining method. To me this has proven to be very effective to filter important information and concentrate on keywords. You can even further enhance this by summarising flows and using arrows to create connections. The usage of drawings and abbreviations can improve your note taking even more. While you take a note, however, you should not forget to keep eye contact with the speaker from time to time.

Take notes to get started

To get started in the project, you should identify individuals that can provide you access to the infrastructure, wiki, and task tracking board. For project related abbreviations, it might be a good idea to maintain a glossary. Once you get access to the project glossary, you can update your entries to it. By looking at the task tracking board or the wiki and by talking to stakeholders, you can gather valuable information about the Use Cases of the software your team works.

Take notes to advance

Once you got started in the project, you can gather further information so you stand out in the project very quickly.

Some tasks that you perform can occur more than once. If that is the case, your team might benefit from notes of the single steps. This also helps you to dive into the project faster and remove impediments in the project as an independent member.

Maintaining a stakeholder list and stakeholder matrixwill help you to prioritise their demands. If youcombine their intentions with the stakeholder matrix you will be able to steer the project better.

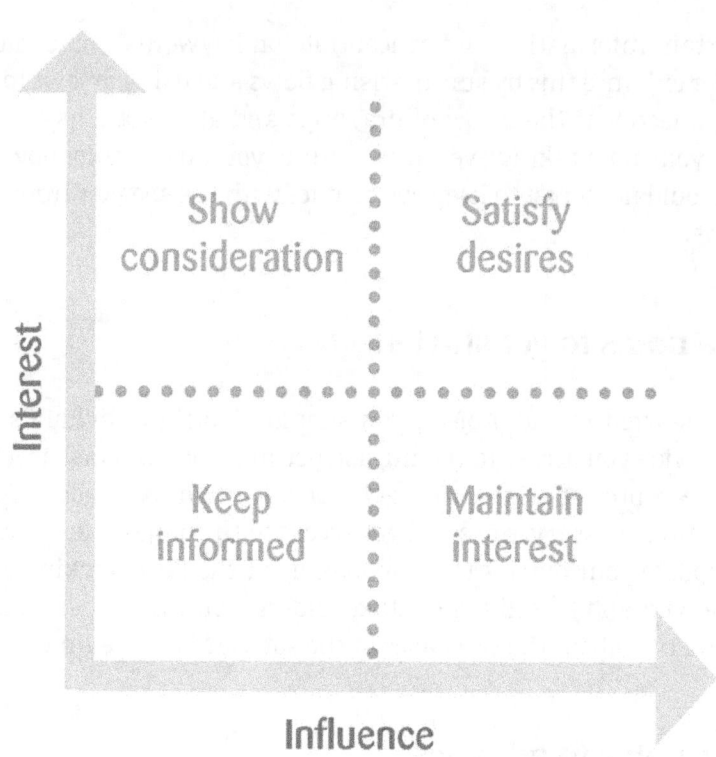

- It is crucial that you understand the vision of the project. The vision consists of a compelling destination, a strategic roadmap, and aligned partners. It helps you to identify the ideal end state of your project. The strategies of the project roadmap should be your base to make decisions that are aligned with the project's vision.

Considering the team's experience is very important. They have a concept of how to steer the project and how to avoid mistakes that have been made before. However, you also should gather enough information to make your own informed decisions.

You should keep a record of your ideas, preferably in a Kanban board. To evaluate your ideas, you can calculate their return of investment and alignment with the project's vision. To prioritise your ideas you should consider whether and how much you can leverage them for your project's current situation. Also it is a good ideal to bring up your ideas in meetings or talk with your colleagues and challenge them together.

Conclusion

When you join a new project, you have to deal with various types of information related to this specific project. You have to be able to extract the important bits and pieces by interacting with the team. Then you have to take notes of this information and store it, either handwritten or on your laptop. By doing this in an organised way with structured notes, you can quickly get to a point where you thrive and increase your performance. In no time, you will be valued as a fully featured team member and you can participate on important decisions.

By Sven Bayer

If you like it then you shouldn't put some code in it

In 2017, a crowdfunding campaign from Australia claimed that "in a world of technical overload there is sometimes no space to fit the simple things on your [bicycle] handlebars." They wanted AU$85,000 to manufacture 'the simple bike bell'.

The 'simple' bike bell worked like this: Instead of striking an actual clapper against a metal bell, you have a small button which you mount on your handlebars. You pair the button with your phone (which must run a supported operating system.) When you want to use the bell, you press the Bluetooth button. This sends a message to your phone (hopefully charged, within range, with Bluetooth turned on, and with the manufacturer's app open and running.) Provided these conditions are all met, your phone's speaker plays the sound of a bicycle bell... probably. What could be simpler than that? (Kickstarter users only pledged 1.7% of its funding target.)

What are the thought processes that lead to trashy or tasteless IoT solutions like this? It's simple: someone somewhere sees an itch, which they scratch with what they know: technology. They then pat themselves on the back for being so 'disruptive', make a few press releases, and voilà: profit. (Maybe.)

Meanwhile, on our morning coffee break, we read the news and roar with laughter. "Remember that awful Fitbit for dogs thing? They were running a six-year-old version of Android. Now they've all been pwned and are being used to mine bitcoin! Hah! What were they thinking?"

We then walk back into the office, and proceed to make exactly the same mistakes as the people who make these terrible 'connected' cat litter trays and 'smart' toasters. As software developers, we follow the same thought process every day. We see someone with a problem, dig into our toolbox of technical solutions, and try to sell one of them to fix it in a clumsy way.

Alarm bells should ring when you end up changing a problem to fit your proposed solution. "The Latin alphabet was created thousands of years ago and is optimised for writing, not reading," says the website for a 'horizontally compact' font. "About time for an update, no?"

The 'update'? Replace the alphabet with dots bearing no resemblance to the conventional alphabet, and squidge words into something that looks like the read-out from a radio telescope. Sound confusing? Not to worry: the website claims it takes "about 20 minutes to get [the glyphs] into your short-term memory. Skip the next episode of Glee and test your mental acuity!"

Where's the joy in reading a good book, if understanding the words becomes a test of mental acuity? Where's the convenience in cycling to the shops for a loaf of bread, if you have to spend five minutes un-pairing and re-pairing your phone with the 'simple' bike bell first?

In 1927, ninety years before the silly bike bell, HG Wells, author of *The War of the Worlds*, wrote in the *New York Times*: "I have recently seen the silliest film." He took issue with the film's core principle, that automation creates drudgery: Wells complained it anticipated "not unemployment, but drudge employment, which is [...] passing away." The film Wells was eviscerating? Fritz Lang's *Metropolis*, in which armies of workers toil in servitude to giant automata.

Seventy-one years later, low-paid armies of workers are employed by financial institutions, to manually enter data into five separate systems; and by social networks, to manually decide if images are too gory, or pornographic, to be allowed. Automation is here, and our half-baked idea of 'efficiency' has begotten drudgery.

The Agile Manifesto defines simplicity as "maximising the amount of work not done." We take this to mean minimising lines of code, cyclomatic complexity, or the number of bugs we fix. Some people even use it to legitimise hacks, or "smashing it in." I think this misses the point.

Instead, we could prioritise minimising the amount of drudgery we create with our software. We could minimise the number of hacks or shortcuts our users have to use to get the result they want. And yes, sometimes this means maximising the amount of work *we* don't do, by minimising the amount of software we build to solve the problem. After all, software means bugs; as soon as it connects to the internet, it becomes an attack surface.

Maximising the amount of stuff *not done* isn't a radical concept. Mechanical engineers know how important it is to reduce moving parts to a minimum. A bike bell that consists of a clapper striking a metal surface is less complex than a Bluetooth button/mobile app combination, and less expensive; it's also more reliable, and more predictable. "Less is more" is nothing new.

We can, and should, do better than to trammel people with unnecessary new drudgery. Think about this the next time you want to embark on a massive, all-encompassing refactor of an old monolith into hundreds of microservices. Take a step back, and don't waste your time on an 'engineer's solution.'

By Jonathan Rothwell

Never forget

We code software. Every day, every hour, an endless stream of code gets written and merged into products. The human-written software codebase is massive. The Space Shuttle runs on 400 thousand lines of code. The Large Hadron Collider uses 50 million lines. And all Google services combined run on 2 billion lines of code.

Software runs technology. Technology has become omnipresent in our lives; from pocket devices to augmented reality and artificial intelligence. The impact that technology is making on human lives is undeniable and inevitable. This fact has its burden: is the technology growing in the right direction?

The answer to that question echoes from the past: it can be found in the thoughts of the first computer engineers and among the ideas of the first technology visionaries. They all promote the very same message: the purpose of technology is not about having everyone interacting online all the time; technology is not a universal remedy (hard) to swallow.

Technology is the challenge for humankind to evolve. It is an opportunity to dramatically increase the collective knowledge, to address the most challenging problems. It is a call to action for public and private sectors to recognize the exponential growth of humankind's challenges, and to provide the vigorous, proactive, strategic pursuit of meaningful evolution.

We, the developers, are makers, creators. We are given the tools and the power to produce the code that will shape the future. We must come up with disruptive ideas that will lead to organisational and societal transformations. Such an attitude should be part of the DNA of any software company that shapes products, services and work. We are here not to code, but to answer the challenges.

Hello world. Never forget to keep evolving.

By Igor Spasić

Pragmatic documentation

Writing documentation is a task that most programmers find tiresome. The pattern I see is that they are asked to write documentation without understanding:

- for whom the documentation is being written
- why the documentation is being written
- how the documentation should be written
- where the documentation is written

By asking yourself these questions, the documentation can end up having a completely different form than originally expected.

For whom is the documentation being written?

With regard to the first question – *For whom am I writing this documentation?* – you should have a clear answer before you even start writing. Sometimes these are particular people you know, sometimes they are existing users that you don't know, and sometimes it is not yet known who the readers will be.

If you have access to at least some of the people who will read your documentation, talk to them about the documentation you are writing. Explain to them that you are doing something they should use and that you want to do it in the best possible way. They will surely help you, because it is in their interest. After all, they are the ones who will eventually read the documentation.

If your users are not yet known, it's best to try to imagine a future reader and try to put yourself in his/her shoes. It's not always easy, but thinking from the perspective of other readers – albeit fictitious – will help you decide what to write.

Why is the documentation being written?

To the second question – *Why?* – the answer is: to explain to the reader things they don't know. Remember this. If you explain to the reader what they already know, your documentation will be boring and will not be read. It is therefore important that you understand what the reader does not know before starting to write the documentation. If the intended audience have varying degrees of background knowledge, this makes things a bit more complicated. You need to know which things most of your readers don't know, and which only certain individuals don't, and organize the documentation accordingly.

How should the documentation be written?

When you have answers to the questions for whom and why you are writing documentation, the third question – *How to write it?* – can be a little easier to answer. Put yourself in the role of a person who needs help. Why? Because, people value their own time. They try to be efficient, and prefer to switch to action as soon as possible, instead of reading up front about what to do and how to do it. Therefore, most of the people won't even read the documentation until they need help. Try to imagine a very concrete thing that your reader might fail to do or fail to understand by themselves. If you already have potential future readers, you can create a list of questions for which they would like to have an answer. It is good if the questions are very specific, e.g.:

1. How do I add a new user who is simultaneously in Role A and Role B?

2. Does the magnifying glass mean a search of data or a detailed overview?
3. How do I recover the object that I just deleted?

If the short answers to the collected questions are sufficient for your readers, then write them down, and call the documents FAQ or Q&A or something like that.

Sometimes, the gathered questions might indicate that there is something that could be done better regarding the user interface. For the three questions stated above, you could, instead of writing answers, actually create UI changes that will naturally lead the user to the answer (a principle known in UI design as "Don't make me think"), e.g.:

1. Instead of the label "Role", use "Role(s)", and when the user adds one role, immediately display a plus sign (+) to indicate that more roles can be added.
2. Add "Search" as tooltip text to the magnifier glass icon.
3. After deletion, open an (unintrusive) infobox informing the user of the procedure for undeleting an object.

Often, through the questions of future readers, you will understand that the problem is that the reader lacks some knowledge to understand your answer. In that case it will be necessary to help them acquire that knowledge. One way is to add introductory chapters in which you explain the necessary concepts. Other way is to provide links in your documentation where the reader can supplement their knowledge.

When you have a text in which you explain the concepts, links to external materials, and the answers to questions that are troubling your readers, you should use this material to group and organize documentation so that it has a logical and easily understandable structure.

As with programming, try to adhere to the principle of least surprise, that is, try to put each chapter in the place where most of the readers would expect to find it.

Where is the documentation written?

The final question: *Where should you publish the documentation?* Answer: Where most readers will look first. Help to log on to the system should be located as a link on the Login page. Help for compiling and running an application can be a file (for example, RUNNING.md) in the root of your project. Help for certain functionality of your software is best if it can be available in the user interface where this functionality is (see previously given example for undelete), and HowTo article for DevOps – on the corporate wiki / knowledge base.

So, the methodology is always the same – putting yourself in the reader's shoes. It is a powerful technique that can be used in various business and private situations, and is especially useful when writing documentation.

By Ognjen Blagojević

Rules are dangerous

Rules followed blindly in a dogmatic way do a lot of harm. This is an uncomfortable truth especially for us, as engineers. As such we are used to looking at the world in terms of rigid rules, true and false, if and else. We like to have crystal clear and exact definitions, guidelines or best practices on how to do something. This is not a bad thing in itself. But we all must be very careful that we do not fall into the trap of just following such rules blindly without giving careful thought to consequences.

Examples of rules from software projects

Here are some examples of rules in software engineering that I have seen to cause troubles in projects:

- The famous rule "working software over comprehensive documentation" from the agile manifesto is often used as an excuse not to write any documentation and deny putting any effort into it. In the long run this is not very helpful in the context of a complex software system where a minimal and helpful documentation is needed to efficiently maintain, extend or operate these systems. Not to write any documentation at all was probably not the intention of the authors of the agile manifesto.
- The dependency inversion principle says "Abstractions should not depend on details. Details should depend on abstractions". I have seen projects that tried to follow this principle very strictly and everywhere in their software. They therefore totally avoided directly using any concrete class as much as possible. As a result, there were a lot of interfaces with only one implementation class each and many factories to create instances of those classes. This

resulted in a code base that is much more complex than needed. It makes the software difficult to understand and maintain without bringing any benefit of really needed flexibility into that software.
- People read the book about "Clean Code" that says that many code comments can be avoided by writing self-explaining code instead. While this is true for many cases, the rule is often misused by developers for arguing that they do not need to write any comments at all in the source code. But there are many things that you cannot explain by using good method or variable names only, e.g. why you had to implement something like that. For this purpose, good comments are still very valuable and should not be denied. I would say that a well-expressed comment can explain much more than a hundred good variable names.
- An article by Martin Fowler about the "Page Object Pattern" stated that one should never put assertions into page objects. Someone read that article and wanted to convince the whole project team that it was very important to follow this rule very strictly. The arguments for the rule in the blog post were not very convincing to me. I brought up many examples where breaking that rule caused much more benefit than sticking to it. Nevertheless, I was not taken seriously. Because it was written by Martin Fowler, it counted as if it was written in the bible. Nobody wanted to constructively argue about it anymore. Fun fact: in the meantime, Martin Fowler changed his article to mention both opinions on this topic. This at least leaves the option open to the reader to decide what works best in their concrete context. Unfortunately, not all famous technical opinion-makers are as wise as Mr. Fowler.

Please, do not get me wrong: I like the Agile Manifesto, the SOLID Principles, Clean Code, Martin Fowler in general, and even his article about the Page Object Pattern. Those are all practices that I appreciate a great deal and that have influenced my way of thinking and working. But, nevertheless, all these things also come with the danger of undesired side-effects, if they are just followed blindly in a dogmatic manner.

The problems with rules

The problems I often see in projects with such rules are as follows:

- Applying rules in the wrong context might not have the desired effect at all
- Using rules more strictly than they were intended to be used causes adverse effects
- Applying rules without really understanding them can result in more harm than good to the project
- Treating rules as God-given, causes people not to reason about these rules anymore. This can cause a destructive discussion culture and lead to bad decisions without real arguments.

Recommendations

Here are my recommendations on how to mitigate the problems with such rules in your projects:

- We must all be always willing to explain ourselves, why we do something in a particular way and what the arguments are for applying a rule in a particular context. Something like "just because our process manual says so" or "because I've read it in that article" is usually not a good answer.
- Don't try to replace critical thinking and adaptive methods with exact and rigid rules that people must follow strictly.
- Do not read my recommendations here in a dogmatic way. Doing so would be to conclude: "Rolf said rules are bad and harmful, so let's

trash all the rules in our project and not follow the Scrum Guide, our Definition of Done, Clean Code and similar things anymore". This was not my message at all!
- Instead: Discuss, write down and live these rules in your team, but see them more as a guide than as precise rules to be 100% rigidly followed in every case.
- Note down not just the rules you use in your project, but also the arguments why a rule should be applied.
- Be open to discussing and re-explaining rules in your projects with your team members and to adjusting the rules to changes in the context (e.g. growing software, growing team, shorter release cycles, different team members, new ideas, etc.)
- Be critical of existing rules, but also respect and try to understand them in the first place.

Conclusion

There are a lot of clever rules or ideas out there on the World Wide Web, in books, in courses or in your projects, as well as in this collection of articles by my brilliant co-workers. Such rules can help you when applied wisely in the right context. But don't let your teams suffer from dictatorship by dogmatic rules and therefore follow my recommendations.

By Rolf Bruderer

Further reading:

Start using a time management technique today

So you take a task from the board, work on it from A to Z and on nothing else. Sounds familiar? I mainly know this situation theoretically. In real life there is always some interruption like a review to be conducted, a bug to be analysed or a phone call to be answered. And having your smartphone laying around vibrating with the latest funny pictures on WhatsApp doesn't help either. Even though we use a prioritised list of work items in our teams, my work as a developer is far from being focused on a single task.

Unlike people who perform at their best when being a "firefighter", I do not like this mode of work. But what can you do about it? Most teams use a regular meeting to improve the way that the team works together. Some things can be achieved here, but on a personal level, you will still have to find a way to focus and deal with interruptions. This is where personal time management techniques come in.

There are a lot of techniques available – I recommend having a look at the following three:

Getting things done

Getting things done provides a comprehensive way of sorting all incoming stuff into various buckets like a waiting list, a calendar, an archive or even the waste bin. I never feel that I have huge amounts of stuff to organise so I have never applied this technique. Reading through the documentation gives you some good hints regarding how to handle incoming stuff though:

- Is the stuff actionable at all? If not, get rid of it.
- What is the concrete next action?
- Can it be done in two minutes? If yes, do it right away.

Personal Kanban

Personal Kanban is an adoption of the Kanban method for your personal use. You basically visualise all your work on the board, limit the work in progress and focus on the tasks you have chosen. I really like having a Personal Kanban board in our kitchen at home, although it is more of a family Kanban there.

Pomodoro technique

The pomodoro technique is, as you can guess, my favourite. It takes a "pomodoro" as a unit of work. Pomodoro is Italian for tomato and the technique's name is inspired by tomato-shaped kitchen timers.

The technique consists – among other things – of five steps:

1. Plan your work: What will you do today? What is your focus during the next pomodoro?
2. Start the pomodoro, for example by setting a kitchen timer to 25 minutes.
3. Work on a single task you planned until the pomodoro (25 minutes) is over.
4. Mark an uninterrupted pomodoro with an X.
5. Take a break (5 minutes after every pomodoro and 20 minutes after 4 pomodoros).

The pomodoro technique helps your brain to be as efficient as possible:

- By marking a pomodoro as done you get a reward for your work. Every 25 minutes!
- Having a break every now and then allows your brain to take a step back and probably come up with new ideas.
- A break also helps you get out of a flow, in which you probably no longer see the whole picture.
- Having a rhythm frees your brain from the need to structure your time.
- By handling interruptions actively, you gain more capacity to focus on your work.
- With the mindset of "having some time at your disposal" instead of "having to fight against the time" you have a positive attitude towards time.

The first thing I noticed after applying the pomodoro technique was that I was more relaxed after work. Only at second glance did I also notice that I was more focused during the work. This is mainly because of the different way in which I handle interruptions when using this technique. Internal interruptions like other tasks coming to my mind during work are just written to the work log – and can be forgotten again. External interruptions like phone calls, emails or colleagues coming to your desk need a different approach. The easy part is all the electronic stuff: switch off Outlook notifications, put your chat to "not available" and your turn smartphone to silent mode – and voilà, you will not be interrupted by these means during your pomodoro. But do not forget to check your email and phone in the next break! Handling working with colleagues is more difficult in this regard: I think collaboration is so important that I generally do not consider a colleague coming to my desk to be an interruption. But at times when I really want to focus on my pomodoro, rescheduling the conversation usually works fine. I just say: "I'll be with you in 10 minutes", make a note and then keep my promise.

It doesn't matter whether you choose an existing personal time management technique, combine aspects of them or come up with something completely different yourself. I just think it is important that you find a way how to stay focused.

By Christian Abegg

Links:

Why you need to go visual

This article is about why you should sketch more in your daily work and why you should use the power of shapes and colours in your documentation. Illustrations allow you to transfer knowledge in a way that is quicker and easier to remember and help to prevent misunderstandings. For instance, if you want to describe the shape of Switzerland in words, you will probably need hundreds of words. The same information in the form of an illustration can be processed by our brain in less than a second and it would still be more precise than the description.

Our brain handles pictures very well

The ability to see and recognise patterns and movement is millions of years old and has improved over time through evolution. Our ancestors' lives were hugely dependent on it. Writing is a human construct for recording spoken words and is only a few thousand years old. This means that our brain handles pictures much faster than text. In addition, it is also more fun to look at and study pictures than to read text. As an example, let's take an IKEA manual of how to build a wardrobe. The illustration-based manual is more effective than a text-based description could ever be.

Easier to understand	Easier to map to the real world	Easier to prevent errors
Less effort to study the pictures than to read a description. Therefore, it's much more likely that people will understand the picture manual.	Easier to map the manual to the real world. E.g. which screw has to be used.	Easier to prevent errors. E.g. the picture clarifies which screw to use. It's less likely that someone will use the wrong screw.

Illustrations in software development

The architecture of software is not that different from the architecture of a wardrobe or another piece of furniture. Both consist of a lot of smaller parts where each small part has to fulfil its purpose. That's why if you use illustrations in your software documentation, each of the advantages from the table in the previous chapter will apply as well. The difference between software architecture and a piece of furniture is that the software architecture is more complex. It has more layers and different life cycle states. Depending on what you want to describe, you will need a different kind of illustration. The following table shows some of the most common diagrams used in software development documentation.

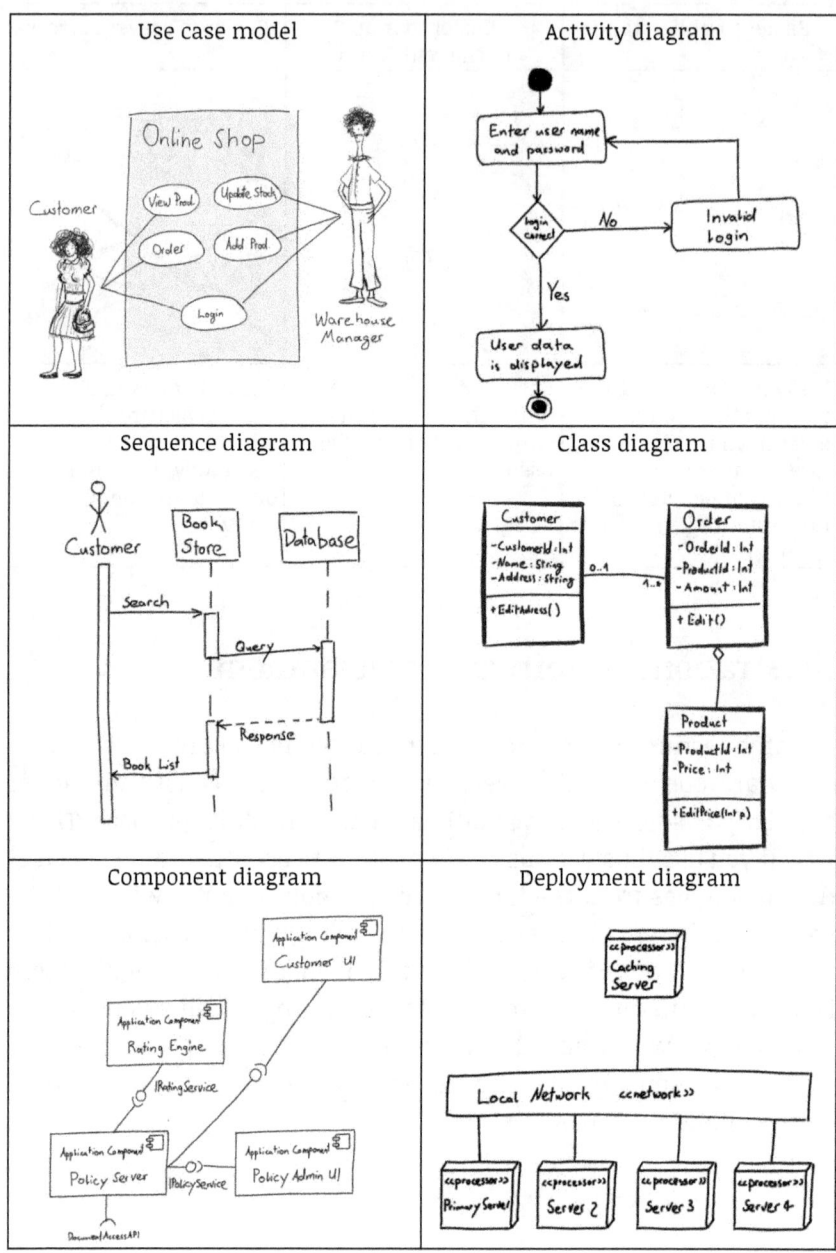

Sketch to understand

If you work on a software development project, you will probably face complex problems on a daily basis. Some common ones are: concurrency issues, async data flow, evolutionary architecture, communication to other applications or how to implement a specific part of the business logic. Sketches can help you to understand and solve many of these issues. Let's take an example. You're a software engineer and have to implement a part of the business logic. A service that creates an overview of all your bank accounts and their credit. The requirement for the transformation will probably be a list of rules, including how to handle different currencies and their exchange rates. Drawing a workflow diagram and having it on paper in front of you will help you to understand the data flow and the connection between the rules. It will help you to simplify your workflow diagram. Maybe some conditions are already included in other ones and can be pruned away. Or you can simplify the workflow by changing the order in which the rules are applied. Manual sketches are first and foremost for you. They should help you to understand a part of your software. So don't be afraid of making a mistake and crossing out part of your sketch. In comparison to a modelling tool, manual sketching is often much faster and gives you a higher degree of freedom. Your sketches can also be used for the system documentation. However, if you are illustrating the architecture of a piece of software that is still evolving, it is recommended to use a tool. This is because it's much easier to adapt your illustration to new changes.

Conclusion

Visualising your work can save you time and energy and sharpen your thoughts. All you need to do is to grab a pen and paper. Try to sketch more often during your daily work. And next time you have to explain your system architecture to a new team member, sketch it.

By Gabriel Duss

"You are not a software engineer" – What am I then?

The job of software engineer, or any technical role for that matter, is extensive. You might spend more time than you would like on meetings, you might spend your afternoon sending emails to your stakeholders, you might actually only get to write code for a couple of hours a day. Depending on your personality type, this can make you thrilled or make you unhappy. I am writing this for the latter. In this article, I offer a different view, in which our job is solving business problems and making our customers happy, while coding is only a marginal tool in this context. I have perhaps hit a nerve with this one, but this mindset is even more important when working as a service provider. So please take a moment to listen.

"But I am a coder, I don't want to deal with any of this."

Yes, I understand and I am too. Coding is fun, coding challenges us, coding is the reason most of us embarked on this journey. But, if you think about it, code is made primarily for human consumption. Yes, machines need to understand it as well, but if it wasn't for the people, we would still write in assembly, or binaries even. When you throw in daily collaboration with your team, managers, and clients, your job is really about dealing with people, not computers.

Let me put it this way. You may put in the work and produce 10,000 lines of the cleanest, most optimised code there possibly is. But if it doesn't bring value to the business, you might as well throw it away.

What is your drive?

You don't want to just play around and not produce anything of worth to someone, I hope.

According to recent models of motivation ("Drive" by Daniel Pink), our actions are fuelled by autonomy, mastery, and purpose.

In healthy working environments, autonomy is easy. You have the power to decide how to implement the solution, to choose the way of working that enables you to do this, and to influence decisions.

Mastery is the urge to improve your skill set and be proud of all the work you have done. This motivates us. To change projects, gather new experiences and work with new technologies. Since the field of IT is a fast-moving freight train, I assume you are already doing this.

Purpose can be tricky, though. The desire to do something that is of meaning and importance, greater than ourselves. Maybe helping an already successful enterprise earn 5% net profit more than in the previous year is not your thing. Maybe you want to help humanity find the cure for cancer. Boarding on some ideas is easy, while for others it is not. I will not provide an answer to this. It is entirely up to you. I will give you a hint though. It is everywhere. It involves feelings.

"I am in the business of creating value."

If your code fails to bring value, whose fault is that?

Remember the cost of fixing a bug, depending on the stage of the software development lifecycle. An hour spent in requirements engineering will save you many more down the road. You might use a framework designed to prevent this from happening, prescribing you the points in time when you should do requirements engineering. But, all the magical frameworks in the world will not help you if you don't understand what you are building.

"I am in the business of satisfying customers."

If you consider yourself to be a one-man business, your customer is every person you interact with. Your customers are your team members, management, stakeholders, users. Keeping them happy is of importance for your business, even if you need to make a couple of trade-offs. Sometimes, you might feel disconnected from the customer. Either he is really busy and doesn't come by, or you see him perhaps once, at a release review. Establishing that link creates a fertile ground for collaboration. Having it is your right, and responsibility.

Anybody can code

After all, we live in a world where anybody can code.

> *"Have you heard? The other supplier was X times cheaper".*

All over the world, you can find coders capable of writing code. And doing it cheaply!

Why would the customer choose you, over another provider?

> *"We use sound design principles, and our architecture is flawless!"*

Yes, kudos for that, you should really maintain that. Selling your work like that totally makes sense – if all your customers are engineers.

To others, that doesn't mean much. The fact that we are proud of our work is not a good selling point – considering the end result is the same: working software.

Own the market

As competing on price is not an option, we might just have a few other tricks up our sleeve. If the customer enjoys working with you because you communicate more clearly, you understand his business and you are actively helping him in making better decisions for his business; these are the skills that make a difference.

By Stefan Djelekar

PART II:
PEOPLE

"As long as you live under my roof, you'll do as I say" – If the project manager leads differently than I would

Here are three observed stories relating to how to deal with personal conflicts within projects.

A colleague of mine moved from working solely with Zühlke people to a project team with members from several service providers. After a few days, he complained that the way of working and the approach in his new project was completely different compared to what he was used to when working with Zühlke teams. He was worried that this approach would not result in a successful project.

Another colleague switched from several Zühlke-led projects to another project in which he was the only Zühlke employee. After one month, during which procedures in the project seemed to change arbitrarily on a daily basis and with non-existent internal communication, he was frustrated and wanted to leave the project.

I joined and assisted this project team during a transitional period of 2 months. After just a few days, it became clear to me that I would not be able to work efficiently in this project, because the leadership and communication were very different to how I would have set up and managed the project.

These are three real examples from my close environment and three times the question is: Is it bad just because it is different and how do I best deal with these "differences"?

To use the words of Henry Ford, I see three ways regarding how to act: "Love it, leave it, or change it"

The most gallant way to deal with the situation is to simply accept it. Which means not only accepting but also trying to understand and then support the way of working, as well as perhaps adopting a different mindset and changing your own behaviour so that it is more aligned with that of the other members of the project. The advantage with this approach is that it causes little friction within the project. But the most important aspect in my opinion is that it gives you the opportunity to try different ways of working and new approaches, and the possibility to learn and increase your personal level of growth, as you find yourself on a new path that you would maybe never have chosen yourself, and can thus gain further experience. A key point from my point of view is that you should remain true to yourself and show your integrity at all times.

If the situation changes in a way that means you are no longer able to deal with it, it won't help to just talk about it in a negative way. Then it is time for a change. Change starts with each one of us making small compromises. The way we share or ask for information, talk to each other, act in meetings and support team members or maintain our network. By enforcing change, you have the possibility to bring the project to the next level of maturity. But there is also a certain risk that you force a change that is not appropriate just because you are used to working in that way. And you have to keep in mind that most people are not enthusiastic about hearing someone questioning their way of working and insisting that they change something themselves. In the worst case, they may try to make the situation worse for you than before, by reducing your involvement and so on. You need to be prepared for such outcomes. Therefore, I suggest handling people with care and involving them in the flow of change, by showing them what they can gain and why they will benefit from the change.

You may have tried to adapt to your team's way of working but then discovered that you are still dissatisfied. And you may have tried to change the way things are done, but instead of finding a solution that is suitable for everyone, you simply found rejection and resistance. At this point, the best alternative is to withdraw or to at least ask the line manager for help with leaving the project. Not as an escape but as a way of protecting you and the team, if the situation is not acceptable and proposals are not welcome.

At the end of the day, you have to look in the mirror and decide for yourself about the best course of action. Was the behaviour of the others (team or project manager) fundamentally wrong or just different? Was I too set in my ways to properly get involved in the new approach or did I try my best to be efficient in the project? Were my suggestions for change appropriate or did I offend my colleagues rather than motivate them? All these questions can help with the decision. – love, change or leave – to be at peace with yourself. The most important point is to take a decision, rather than just creating a negative atmosphere in the team by e.g. taking negative actions or making negative speeches. Because a negative atmosphere is the worst thing that can happen in a team, and will result in it losing its effectiveness.

By Sabrina Lange

Business readiness – is there readiness for agile development in business?

The development departments are increasingly aligning themselves with the agile topic. As a result, agile development is gradually becoming state-of-the-art. Time and again it creeps into the interface to "Business", as "the other" part of the company is often affectionately called. This is because Business should be closely involved in the development work, as a Product Owner (PO) or a Product Manager (PM à la SAFe), in order to ensure that business value and customer benefits remain in focus and have a high priority. So members of the development department ask: "Is Business ready for agile?"

When tidying up my mailbox, I came across a link to a SAFe discussion in 2016: Under the title "Business readiness for SAFe", someone asked the group for material about the benefits of agility (or SAFe) from a high-level business perspective. The aim was to identify clear benefits from a business perspective, rather than pure development benefits: "Not how effective it is to have cross-functional teams, or why Scrum is great". Brilliantly put! Exactly, such aspects are of no interest to Business. So what did the group have to offer as answers? Curiously, I read on.

And I was rewarded with a wonderful response from a member of the forum: "Agile is much closer to how non-IT business functional areas work. What is a 'new way of working' for us is how they've always worked: collaboratively and in response to a constantly changing environment". Many thanks, Janet. She provides the example of a legal department that does not have to write a PMO or a change request to adapt to new regulations. "The IT community has constructed its own cage".

This is a tough but clear formulation. We old hands remember: the project management that we now lovingly call "waterfall" had its roots in IT or development in many companies. That was where the first experts who knew how to systematically use the approach were located, and from there it was conveyed to and promoted in the company, in the specialist departments, until ultimately there were business project leaders and even pure business projects. For "scaling", there were various boards at the program and portfolio levels, to help manage the interests and conflicts. IT/Development is now ridding itself of this "corset" thanks to the "agile revolution" and is asking: is Business ready for agility?

It does not matter to me if this assumption is actually correct. What I like about the wording is the change of perspective: it is not about good IT/Development against evil, unwilling specialist departments, nor the other way around. The reason for the company environment currently working the way it does is not (just) because it has been prescribed that way by the specialist departments or management, but because it has been actively shaped and cultivated in that way. Dear readers: please remember this the next time you are irritated with the specialist departments.

What do we learn from this? As a company, IT, Development and the specialist departments are all in the same boat. None of them can solve all the problems on their own – so let's do it together! Have you ever wondered how the specialist department works? How does it handle decentralised decisions and transparency? What makes its world so different from yours? Be curious! Don't be afraid to ask.

A note from me: I came across similar topics in relation to Business Process Management. In this context, the topic of business readiness and change management keeps coming up. Here there are Process Owners, who are a bit like the Scrum Master, as well as the conflict between process and line organisation, with which we are also familiar in project management in Development.

I have been building bridges between IT/Development and Business for 20 years. The distance between them remains, as the differences in language and visions are too great. But it is precisely these different perspectives that provide enrichment and help us move forwards. Remain curious. Help those who try to build bridges. Agility, SAFe and Lean will not be sufficient on their own, as they simply provide a modified way of building bridges. Ultimately, people have to work together and communicate with each other for the company to succeed. And, as always, respect, empathy, openness and transparency help in this regard. And respect. And reliability. And respect. But that was the same in Waterfall times as it is in Scrum times.

Remain curious!

By Ina Paschen

CYA: Cover your ass

Imagine on a Monday morning you come into the office, start up your computer and ba-bam a manager is standing beside you, telling you to follow him into a escalation meeting.

There is a problem with the software you are building in production. In the meeting there is the CIO, your line manager, the technology line manager, the operation line manager, the central architect line manager and you.

There is a problem with a library you introduced into your software a year ago, which is causing crashes of the software in production and the company is losing money because the users cannot work.

The managers want to know why you have chosen to use this library and who gave you the sign off to use this library. There are now two possibilities:

A) You start like hmmmm, this was a year ago and hmmm, actually I don't know, but this is not of interest now, let me go and fix the problem … very bad idea! You're screwed…
B) You take the laptop and navigate to the list of architecture decisions, you show them the architecture decision with the evaluation and the approval. And now you say: "Can you excuse me? I have a problem to fix."

Of course: All characters and events in this article are entirely fictional ;-).

I know you hate to document things and I know that you think you will remember everything. But just take my advice for your career.

Use CYA = Cover your ass. CYA has one simple rule: Document EVERY Decision => DED

Yes, document every decision. Whether you are a Software Architect, Business Analyst, Consultant, Engineer, Developer, Manager or a fluffy unicorn dancing on a rainbow.

So, how do you document a decision?

You create a decision log in a tabular form in a suitable medium (Git, wiki, SharePoint, Word, Excel, ...).

Number: Every Decision has an id or number. Which can be used as a reference.

What: What is the decision that was taken

Why: The reasoning and arguments, constraints, implications and references.

- Context: In this part, we document the context of the decision. We give the reader some extra information, so he/she can understand the context of the decision.
- Problem: The problem or the challenge giving rise to this decision.
- Decision: The decision that is made. Document the evaluation you have done that has led to this decision here.
- Consequences: The consequences of the decision.

When: When was the decision taken?

Who: Who was involved in taking this decision? Hint: The more people who agreed on a decision the better.

No.	What	Why	When	Who
D-1	We will use a decision log in our project	Context: • Decisions need to be documented so that everyone knows why a decision was made. Problem: • Not everyone remembers after more than a year why a decision was made. • Knowledge drain: People leave the project Decision: • We introduce a decision log. Consequences: • Every decision is documented in the decision log	01.01.2020	Hans Muster Simon Stucki

What happens if a decision is revoked? Strikethrough the decision and define a new decision, where you document why decision No. X was revoked.

CYA = Cover your ass will make your life easy and everybody in your team will know where to look to see why a decision was taken, so that you can move on to the next project without being haunted by unknown old decisions from the past.

By Romano Roth

Digitalization and its impact on customer interaction

What are the core principles of digitalization?

There is no doubt that we all are in the middle of a digital transformation. For most companies it is clear that they are affected in some way by digitalization. However, it is less clear to them HOW they are affected and WHAT they should do to prepare for the challenges of the future.

It all began with the launch of the Internet and then seemed to slow down after the dot-com bubble. But, in 2007, it really started with the launch of the iPhone, when it became obvious that computer power had reached pocket size and that "having information at your fingertips" was the new state of the art.

But what exactly has changed? And are there common patterns of digitalization in all the different industries and business areas? What should a company do to remain competitive in the new digital age?

It is important to understand the core principles of digitalization in order to align the actual business with the future:

1. **Global availability and tradability of digital services and products**

We notice many small changes in our daily life, but easily overlook the dramatic changes as a whole: telephone booths have vanished in just a few years; CD shops have disappeared from the high street; kiosks are transforming from paper magazine sellers into digital accessories shops; cinemas are fighting for survival as customers can stream formats without latency and enjoy films wherever they are and whenever they want; advertisements are appearing more and more frequently between our online news bulletins, but disappearing from newspapers (which themselves have a tendency to disappear); augmented reality warns us of traffic jams right when they are happening and interactive street maps with GPS positioning have replaced paper street maps. When you try to think of what types of product you no longer need because you now have them on your mobile, you might struggle to write down the first 20 things, but then you will hardly be able to stop writing.

Even more important is the fact that these digital products are globally tradable and almost immediately available. You just download a book when you first hear about it and seconds later you can read it. A train ticket from Zurich to Stuttgart can be bought in Switzerland or in Germany and you will probably buy it wherever it is faster, cheaper, easier or more convenient to do so.

Today's customers on service websites don't complain very often, but they appreciate a good and friendly service. They want to receive immediate, but personal service and publicly rate the service provider and the user experience on the internet. When a family member was given incorrect medical treatment and struggled to get hold of the doctor, I made the treatment public and, suddenly, we were astonished to discover how easy it was to find a mutually beneficial solution with the hospital, under the condition that the rating would disappear as soon as corrective measures were taken by the hospital. Multilingual trading platforms link sellers and buyers, and much of the traffic on those platforms is made up of electronic online transactions. And if a service is down, customers might change to a different platform within seconds and without hesitation – maybe never to come back again. We realize that, in our digital world, we can lose our clients before we have even met them for the first time. Never have processes been more critically important than today when it comes to satisfying customers!

2. Vertical and horizontal integration among customers, partners and suppliers

Customers, partners and suppliers are often in the same commercial network to profit from "just-in-time" advantages. There are new ecosystems and platforms evolving that offer advantages in logistics (such as cross-docking synergies, delivery services, billing and dunning services and even factoring services). Automation makes it possible to serve millions of global customers digitally, in an easy, friendly and personal manner. Sometimes it just makes sense for the customer to get everything from one source and resellers are frequently no longer required for digitalized goods. Global sellers like Amazon and Apple have learnt to address end users effectively. Some companies realize only too late that they are about to fall out of the process or get into a new dependency because their customer relations are not aligned with their digitalization requirements. Only companies that are focused on the needs and values of their customers will make the change and be prepared to stay independent and act strong.

3. The value of a constant data stream for digital services

The data streams that result not only from customer interaction, transactions and enquiries but also from registered customer habits have become a major source of income and research alike. Not only is Google, for example, able to design customer profiles, but it can also constantly improve its search algorhithms. This allows it to constantly improve its position in comparison to its competitors (that are still trying catch up with their Google search ranking). The advertising market depends on this kind of information in order to optimally personalize the way they address customers. Companies with good access to customers, such as Facebook, Google and Amazon, can innovate new intelligent products based on this data. And, last but not least, the servicing of products can improve if valuable data is available. This data stream is an extremely valuable asset and companies cannot afford to do without it.

There are some products, such as harvesting machines for farmers, that people tend to assume, being heavy machines, will always remain purely physical. But the data stream relating to those physical products can lead to new kinds of after-sales services.

John Deere has videos that illustrate the benefits of digitalization for farmers in connection with IOT concepts. This data might additionally be used for predictive maintenance or might be enhanced with IOT concepts and could enable a whole new type of return for the value chain. The data stream is a competitive advantage that helps with the designing of new digital services, especially in evolving domains.

4. The value of data streams for the physical world

However, digital information might not just be used for digital extension of the service range: While I was writing this article, Amazon announced that it would also use its digital information about physical products in order to optimize physical book stores. At Amazon's six physical stores, books are arranged on shelves face out, even though this takes up more space. Amazon is not trying to cram the entire inventory into these stores; its view is that you can just order everything else from your phone. It also devotes a lot of space to its

Kindle e-readers, streaming TV devices and other gadgets, so you can try them out before buying.

What is the takeaway for companies that are affected by digitalization?

First of all, digitalization is particularly beneficial for platforms that link buyers and sellers in the same ecosystem. If done well, other companies might want to profit from your excellent relationship with your customers. In the long run, it might be more profitable to own and run such an ecosystem than to operate a traditional business.

It is of paramount importance that you put the customers at the centre of all your thoughts. Try to surprise them and provide good feedback facilities in order to constantly learn about customers' needs and concerns. In order to achieve this, processes are no longer sufficient, as they rarely reach beyond the company itself. What you need today are user experience mapping tools that enable you to assess how customers feel about the products and services they receive.

Design the customer interactions from the outside to the inside: Increase the understanding of how (real) customers interact with providers, then design and speed up the interactions from an end2end perspective by eliminating media gaps and by increasing data quality.

Embrace technical advances for the benefit of your customers. New technologies often bring opportunities to improve products or services. The 5th generation mobile services started in 2018, with the first calls, and will be broadly introduced in 2020. They will eliminate the latency from connections, so that cars being driven in traffic can interact in real time and optimize security on the streets. Have you already evaluated how this could also change the just-in-time concepts of your production?

And, last but not least, challenge your business models. Even though your business model has worked so far, it might be possible to add additional value streams, with additional business models or with combined models. A good starting point for this might be Oliver Grassmann's book "The Business Model Navigator: 55 Models That Will Revolutionise Your Business".

By Beat Bischof

Discipline flow

Product conception, development and delivery is a highly interdisciplinary team endeavour. Anybody who has ever participated in a more or less complete product lifecycle knows that not only are the languages each discipline uses quite different, but also the results each discipline obtains may not always be compatible in the manner intended.

The discipline flow tackles this by relating the engineering disciplines to one another in a general product lifecycle context. Its core use case is to get a comprehensive overview of the disciplines and flows required for product conception, development and delivery – from user needs to business goals to the product as delivered by operations to the final user.

The basic model

The very first step of any kind of meaningful product development must take the user as well as the business goals into account and, by this means, allow for continuous delivery of product increments. Relationships between the Business (BUS), Customer Experience (CX), User Experience (UX), Requirements (REQ), Architecture (ARCH), Implementation (IMPL) and Operations (OPS) define the basic flow model to achieve this.

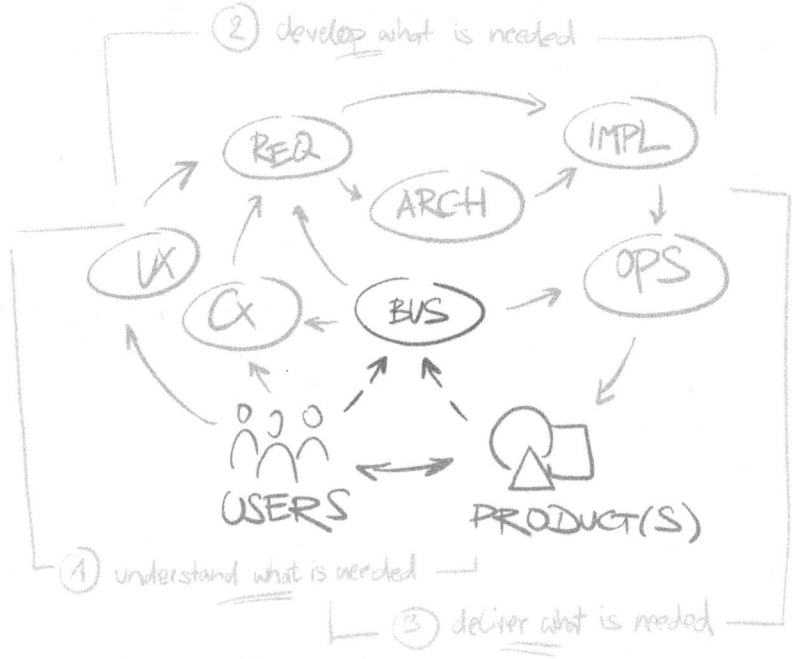

The various disciplines concentrate on discipline-specific essentials:

- BUS: Business, organisational and societal goals for the given product(s)
- CX: Customer's overall journey
- UX: User and his goals
- REQ: Product features and acceptance criteria
- ARCH: Quality attributes and architecture
- IMPL: Components and connectors
- OPS: Infrastructure and monitoring

Additionally, one might think of testing (TST) as a discipline on its own, for instance between IMPL and OPS or OPS and the delivered product. In some project contexts this would make perfect sense; however, in general, we'd prefer to think of testing as an activity that is integrated within each discipline. For instance, in BUS we may think about how to test the business model hypothesis, as part of UX we may think about how to do the user acceptance testing and in ARCH the testing for quality attributes is paramount. Of course, all this needs general testing knowledge as a foundation and this is where testing as a discipline might play an explicit role in your project again.

The different disciplines are grouped into three major parts, each with a different focus on what has to be achieved: (1) Understand what is needed (2) Develop what is needed (3) Deliver what is needed. These three quite different perspectives also provide clues as to the characteristics and attributes needed in your team.

We then monitor the performance and usage of what was delivered, incorporate feedback and start again in order to develop corresponding improvements and extensions.

Discipline experts

The disciplines and relationships described above form the core elements of an abstract product delivery machinery. However, the machinery is only the basic structure. It needs to be implemented by people who are experts in their disciplines. At the same time, they must have general knowledge of all other disciplines in order to be able to always put specific work into the context of the overall product lifecycle. Only then may we avoid silos and waterfall-like delivery processes.

The more directly the disciplines are related, the more this knowledge is required. For instance, working on the architecture requires a solid understanding of all the requirements, both the functional requirements and the requirements for quality attributes. And in the downstream flow we need a thorough understanding of which infrastructure is needed for effective operations. This holds true not only for production, but also for development and test environments.

Project excellence

We need people working together as a team, often facilitated by somebody in the team taking care of collaboration topics such as establishing and customizing an agile process, enabling appropriate levels of self-organisation while also considering individual personal development goals. At Zühlke, this is typically done by a collaboration owner (CO), elsewhere this might be a Scrum Master. Additionally somebody taking care of project budget and risk is needed within the team. Typically, this is done by a team member from the project management domain.

Neither the collaboration owner nor the project manager (PM), are part of the sketch above, since both roles are often more concerned with the complete flow and not so much with discipline-specifics directly, at least in their roles as CO and PM. The PM, of course, often also considers the ROI of the product in relation to the project's budget and hence is part of the BUS discipline. Similarly, the CO might be part of a particular discipline. More importantly from the flow perspective, both roles together ensure that not only are features delivered, but they are also delivered on time and in budget, with appropriate quality, by a team that sees their purpose individually as well as at the level of the product development as a whole.

Successful products

Even if we have achieved project excellence, this does not ensure that we have delivered a *successful* product. These require additionally that user and business needs and goals are explicitly taken into account from early on. We need to develop and deliver incrementally and incorporate feedback and learning continuously. Each discipline has its specific contribution:

- BUS: models with statements about value and impact
- CX: overarching product(s) delivery and experience concept
- UX: statements about product value and specific concepts for user needs
- REQ: concrete requirements and metrics for the product development
- ARCH: feasible technical context for development and production
- IMPL: adaptable, maintainable and operatable implementation of what is needed
- OPS: concrete environments, pipelines and monitoring for development and production

One key aspect in the achievement of product excellence is to track and monitor product features (or increments) as they make their way through the complete cyclical flow in order to learn and improve continuously within and from each discipline. That is all the way from CX/UX to delivery in production systems, including the tracking and monitoring of the product's usage and business performance.

Conclusion

In general, the discipline flow makes the big picture of disciplines and their relationships a first class citizen and makes it possible to explain what we sometimes call 'Operational Excellence'. This kind of excellence is a result of discipline experts working together within a lean project organisation for product delivery that focuses on user and market needs from the very beginning. The flow comprises only a small number of disciplines and relationships, simple enough to sketch ad-hoc whenever needed for discussions and complete enough to check the continuous delivery maturity in the context of your current project.

By Stephan Janisch

Don't teach kids programming

Programming is increasingly being introduced to primary schools. This is an initiative that is recognized all around the world – many kids are being taught programming.

Stop! We're wrong! Do not teach children programming!

The assumption is wrong. What we are doing is observing the present and noticing the rising trend of the need for developers. We extrapolate this fact and base the future on it, assuming that the same rules will apply in 10 or 20 years from now, at the time our children become old enough to work.

If there is something we do not know, it is what the future holds for the world. The dynamics of change in the digital industry are so extensive that there is no pattern which can be applied to them. The amount of information is multiplying; requirements change faster than ever. The truth is that we have no idea what the world will look like in 20 years. In such an environment, programming is, unfortunately, not a "joker" wildcard that will give our heirs a chance to master the world of the future.

Moreover, the type of programming the IT market is looking for is mercilessly monotonous and stumbling. It is all about the skill; programming today has been reduced to being more about the framework timing, and less about the science. Do we really want to involve children in such an anaemic world of programming?

Programming should not have a meaning in itself. Programming should be a tool – in fact, only one of the tools that will be available to people. The technical knowledge which we boast of and so passionately wish to put into young brains should not be taken as the primary source of knowledge.

Instead, we need to teach children critical thinking. In a world without censorship, but with fake news, a critical attitude is more important than programming patterns.

We need to teach children communication. A world in which everyone has a voice and an opinion about everything requires precise and clear communication skills and the ability to exchange ideas and thoughts.

We have to teach children to work together. In a world where there are more screens than people, cooperation becomes a necessary ingredient of progress.

And finally, we have to teach children creativity. Creativity is a part of what it means to be human. Creativity is something we need to constantly stimulate, now more than ever before, because that is the only way our children will discover how the world of tomorrow will function. Do not teach children programming. Teach them that they can and should change the present – that is going to be our future.

By Igor Spasić

From enthusiasm to commercial success

Throughout history, the goal of every human has changed. At the beginning, the sole goal was to survive. With the evolution of mankind, a large percentage of people could work regularly without worrying about survival. Today, our work is not only the source of our income. We want to feel that we are contributing and that we matter. We want our ideas developed. To achieve this, there are various important factors that we need to understand.

In this article, I will provide some hints taken from my personal experience and the experience of my co-workers. These hints can be summarized as follows:

- know your limits
- do it now
- know your priorities
- know your audience
- be ready for failure

Literally everything is achievable. But there is also a price you need to pay to succeed in achieving it. Behind this slogan, there is more than just a motivational message. It also means that it is not only about the effort you put in, but also about how much time and resources you devote to it. And, you should not spend 100% of your day working, because a private life is important for your happiness and can significantly improve the quality of your work.

You should not postpone the start of your journey until "the time is right". The time will never be right. As a matter of fact, it will never be "more right" than now. So, if you want to make something big, the sooner you start, the better.

What you should also have in mind is that it is not only about putting in enough effort. You need to thoroughly divide and plan your work by knowing your priorities. You can base your decisions on something like the 80/20 rule. If you plan and prioritize your work properly, 20 percent of your work will give you 80 percent of the value. Of course, these numbers are not exact, but they can show that the majority of the value is delivered by the initial effort you put into something.

No matter in which area you are working, you should know who your main target group is. Imagine you have created a teleportation machine. If your product can only transport individual particles of dust, no one is going to be interested in that. Additionally, the outcome of your work must be beneficial for the majority of your target audience. So, if your teleportation machine could transport humans, but only from your living room to your kitchen, you would impress many people, but nobody would use your masterpiece.

One of the things that is hardest to accept is the fact that you should be prepared for failure. All the points mentioned above will help you. But, whatever you do, there is no silver bullet for success. You will just have to repeat a try-fail sequence until you finally succeed.

All these points are applicable when you are creating a product. But would they be applicable when exploring a new area, let's say, a new technology? The answer is simple. If you cannot show the result of your work, you will not convince anyone that you are knowledgeable in the area you have explored. To do that, you need to create something that gets people's attention and that shows the benefits of your work.

I would like to give an example by sharing my personal success story. In Zuhlke, we have focus groups. These are groups of people that are interested in a particular area and are putting an effort into exploring it. Some time ago, we started a focus group which explored augmented and virtual reality. We did some investigation and gained the necessary knowledge, but we realized that this was not enough. To show other people our capabilities, we decided to create an app that would give us visibility. We analysed the market and noticed that there were many different shared whiteboard applications, each of them lacking tactile feeling. So, that was our goal. While working on this app, we felt that we were contributing to the world and really cared about this product. The first version we implemented was not even close to final. But it was enough to show that it can be done and that we knew how to do it. Eventually, we were heard and, after some time, we got the commercial project.

There are many other similar examples within our company. What is common to most of them is that they came about as a result of different focus groups. Thanks to these groups, our engineers gained expertise in many different areas. This was helpful for both the company and the engineers themselves. That is why companies should be willing to give some space to their engineers to develop their own ideas.

A journey to success is not a straight line. There is almost no chance that you will simply think of something, make some effort and succeed. However, you can make it easier if you invest your time not only into implementation work, but also into the aspects mentioned above. They might seem less important, but they can significantly reduce your effort and increase your chances of success.

By Bojan Jelaca

Know-how transfer – just explaining once is not enough

Most projects face situations where knowledge needs to be transferred from one person to another. Integrating new team members quickly, with an effective and efficient know-how transfer, minimizes delays and transition costs – you can react flexibly to changing requirements and ideally specialists can contribute to several projects. How can you ensure that nothing of importance is being lost even in complex projects?

Everything, as fast as possible

The expression "know-how transfer" suggests that knowledge can easily be handed over to another person. But it is not that simple:

- Often knowledge is only available implicitly and distributed over several people.
- The time to hand over information and the capacity to adopt it is limited.
- Knowing something does not automatically mean it is understood or can be applied.

How should you organize a handover then?

Set priorities

First and foremost: set priorities. It is an illusion to expect that all knowledge can be entirely transferred. Therefore, a better strategy is to ensure that all central topics are well covered. And remember, priorities depend on the perspective: perhaps the knowledge recipient has a different mission? Surely the recipient is the one that needs to work with the new knowledge and the one handing over is often no longer available after the transfer. Hence the handover process should always be led by the receiver, who therefore needs to know their goals.

It is a legitimate concern that entire topics may be overlooked when following this approach. To avoid large gaps, be sure to start with a high-level overview in order to understand the bigger context. Knowledge maps and user story mapping can be useful tools to support this process.

Documentation

If good documentation is available, the handover can be simplified to providing an introduction and overview of the documentation as well as explanations for undocumented topics. It is crucial, however, that the documentation is trusted and up-to-date. As a minimal structure, on-boarding checklists have proved helpful. These should not just take the form of a list of technical setup tasks, nor that of comprehensive documentation, but should rather consist of a map of the most important topics and where to find documentation. They should be small enough to manage so that they do not become outdated as quickly as (yet another) documentation.

Apply what you have learnt

Documentation provides explicit knowledge – a conscious effort to preserve and make available knowledge about a topic. Much more difficult to share is implicit knowledge, acquired from years of experience.

For instance, knowing with whom to consult about various topics as the result of past interactions with persons within a network.

Typically, the absence of implicit knowledge is first noticed once it is needed – when the "real work" begins. An effective method to ensure that you have the required knowledge is to work under the supervision of the knowledge giver. In software development projects, this could include fixing bugs, implementing new features, creating and executing tests or deployment of the application. To quote Benjamin Franklin:

> Tell me and I forget. Teach me and I may remember. Involve me and I will learn.

Doing practical work in the acquired codebase will immediately reveal the knowledge gaps which need to be filled. More importantly, knowledge learnt is more likely to endure if it is a result of practical work.

Continuous know-how transfer

Even if no employee exchange is planned, it is good practice to share knowledge within the team:

- Exchange sessions covering important topics in order to increase awareness within the team
- A group chat tool with good search functionality to simplify informal exchange, especially within distributed teams
- Methodologies with short feedback loops to foster exchange within the team

The above measures will help to avoid knowledge silos and, at the same time, lessen the impact when a team member leaves.

Check the success

As a knowledge receiver, make sure you understand what you receive. The transfer of knowledge is not a one-way street; ask questions and try to apply what you have learnt. A simple and very effective check to see if you have really got the point is to give a short summary of the topic to outsiders. And give some thought to the next joiner: offer feedback on how the on-boarding process can be optimized.

Without such checks, you might grasp a passive understanding of the topic without acquiring the ability to apply it in practice.

Hopefully, these suggestions will help ensure that your next project change is successful.

By Christoph Zuber

Lean startup: taming the uncertainty

Have you ever been involved in a project with an unclear vision? Maybe a project where the stakeholders disagree about who the product's main users will be? Or a project with a long list of "must-have" features but no agreement on how to prioritize them?

The next time this happens, try the Lean startup methodology to tame your project's uncertainty. Some of its principles are:

- treat everything you believe as an assumption that needs to be tested
- build the smallest possible product increments and use them to get market feedback
- build up your knowledge through many quick iterations of the build-measure-learn cycle
- fail early and then change course based on what you have learnt.

Classic examples:

- Zappos; started not by creating an online store for shoes, but by taking photos of shoes in retail stores and putting them on a static website. The founder processed orders by hand. Goal: learning what customers want from an online shoe retailer.
- You are thinking of creating a newsletter. Instead of hiring a newsletter team, just add a sign-up form on your website. Define beforehand how many sign-ups you want before you actually start producing the newsletter.

"Yeah, nice!" you say. "It's for start-up founders, just as I expected from the name. *So why should I use this method in my development or consulting project?*"

Here's why:

All projects have to cope with uncertainty. It's highest at the outset, but it never goes away. That's why we employ agile methods. Lean Startup is useful because it helps you decide what to do next under conditions of uncertainty: everything is an assumption until you validate it. Find the assumption that carries the biggest risk for your project, then look for the simplest experiment to prove or disprove it.

Do you understand the business model of the product or service you are helping to build? Who are its customers? Which of the customers' problems will the product solve? What is its Unique Value Proposition and how is it going to be implemented (the solution)? To answer these questions, spend one hour with your team to fill in a Lean Canvas – a one-page business model consisting of nine segments. It requires you to spell out all your assumptions about the product and supports a shared understanding of what you're trying to build.

A Lean Canvas can have additional benefits:

- The customer segments point at personas to look into during UX research
- The Unique Value Proposition will make discussions with stakeholders and users more purposeful
- The solution can be a starting point for determining the project's scope and technology
- Prioritizing the risks on the canvas might get your risk analysis off to a flying start

With the Lean Canvas filled out, a first shot at the business model is right in front of you. Now tackle the uncertainty by conducting experiments that challenge the assumptions that carry the biggest risk. Typically, those are the customer segments and the problem to be solved. Two very simple experiments are problem and solution interviews as described by Ash Maurya in "Running Lean". Use them to derive a deep understanding of your customers' problems and to design and validate a solution that fits their needs.

A problem interview is fairly easy to conduct. Identify some potential customers or "prospects" that are willing to discuss your idea for an hour. Tell your prospect a story that highlights the three most important problems your product or service will attempt to solve. Ask the prospect to rank those problems. Then, ask them how they address each of the problems today. At this point, most people will tell you their story. Just sit back, listen and learn about their world view. To deepen your understanding, ask open-ended follow-up questions and, of course, take notes about everything you hear and observe. This procedure will provide you with a wealth of information about your customers' real needs, thereby reducing the uncertainty in your project. Use the insights to update your Lean Canvas. If the problems have been validated, quickly create a demo to test with your prospects in a solution interview. The demo should be lightweight, easy to change and look realistic. Demonstrate to your prospects how your solution will address each of the problems validated previously and ask whether they would use it. Again, sit back, listen and learn. If the prospects respond positively to your demo, test your pricing by telling them what the real price will be.

Performing these steps together with the team does not take much time, but it does a great deal towards taming the uncertainty inherent in your project.

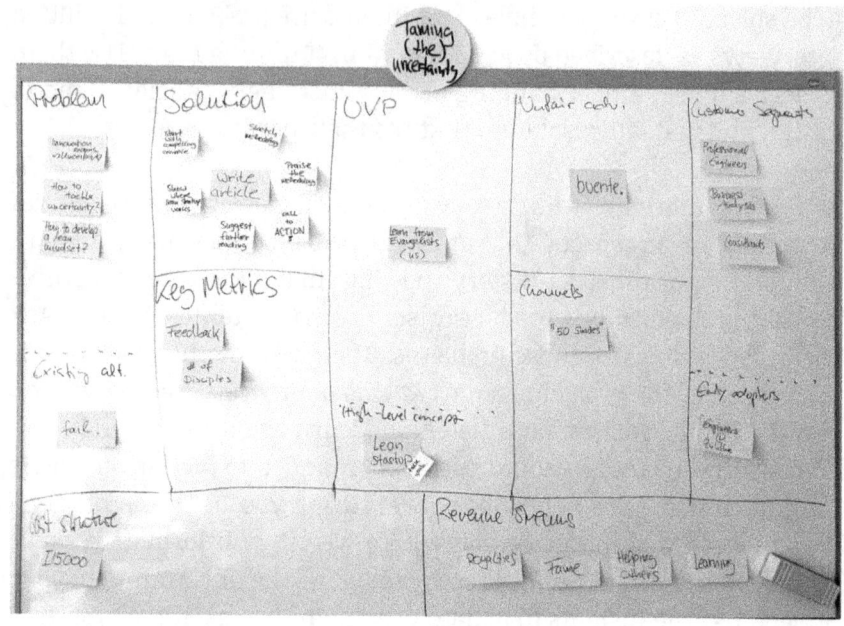

Further reading:
- Eric Ries – The Lean Startup
- Ash Mauria – Running Lean
- Niklas Mådig & Per Åhlström – This is Lean

By *Eric Fehse, Sven Bünte, Stefan Reichert*

Meeting with users is essential for creating great products

Even though – as a UX professional – I must stress the importance of meeting users, I have to admit that the title of this article is not exactly accurate and just meeting users is not really the point. Having revealed this, I should probably explain what really matters and give some indication regarding how to do it.

The purpose of meeting users

Challenge #1: We who develop a technical system are not like the users.

This has two key reasons: (1) The more we are involved in the development, the more elaborate is our mental model about the system and how the system is meant to be used. (2) Even if we are users, we are just some of them. There are usually many more users with quite different needs and mental models.

Conclusion #1: the more we rely on our own evaluation of the product we are creating, the more likely the product is going to suck

Things are even more interesting. What users really need changes constantly. And it is most vexing for some of us – for others it is a fantastic opportunity – that we who create technical systems drive change. Introducing new possibilities changes what people are doing, how they are doing it, what they trying to achieve and thus what they expect from the technology we provide.

Challenge #2: a new product changes what people need from a product.

The better we can anticipate the change, the better our products will meet the future needs. And to anticipate a change we have just one option: we must create the solution, either as a mock or even as a product, and let users use it as realistically as possible. From this, we can learn what the future could hold. By the way, people can get quite creative and do stuff nobody ever thought of doing with our products. If we can harness this creative power, our products will rock the market. The key is to be one step ahead of the users. While they adapt to the new technology, you observe them, create new ideas and start the next cycle.

Conclusion: *the purpose of meeting users is to learn what tomorrow will be.*

How to meet with users

Get inspired by life

Usually, meeting with users starts with life as of today. Go and observe people, talk with them about what they are doing, why, what they love and what they hate. You can also become an apprentice and do it yourself, instructed by a master. Identify their tasks, their values and beliefs, what they try to achieve and what hinders them. The goal: Gain understanding of users and their context, identify and test possible stories.

Things to be aware of:

It's a learning process. So, steer your progress with open questions, analyse each session immediately and rework your open questions for the next session.

Participants will give you numerous solutions. Listen and learn what they really wanted to achieve, see the problem and get creative to find an even better solution.

You need a wide range of opinions so pick a wide variety of participants.

The essence of the problem lies in the daily hassles. Thus, you should experience real work, dreams and everyday life. Avoid talking about generalisations.

Capture the details as direct quotes, and collect pictures, materials and forms that people are using. This material will later allow you to quickly create prototypes that can handle real life problems.

You need to consolidate what you learn from several participants: a big wall and sticky notes do the trick.

Be sensitive to commonalities and differences between the different persons you talked to and work them out. There is no average user.

As a result of such activities, you can expect to have much deeper knowledge about users, their life, their work and their dreams, as well as loads of ideas about how to achieve a dream or two.

Evolve the product story

A product story answers the very fundamental questions of the product. Alas, just filling out a lean canvas and an idea sheet is not good enough. We must probe the users and get some more substantial evidence. Is the problem relevant? Does the solution fit? Is it the best possible solution? There are a range of method frameworks like design thinking, design sprint, user experience sketching, contextual design, lean startup and more to help you. Whatever the name, they all basically describe building a team that iteratively creates and evaluates solutions with users.

A few methods to use here for meeting users:

- UX sketches, physical mock-ups and paper prototypes
- Narratives, storyboards
- MvPs and test implementations.
- Hallway testing, UX walkthroughs
- Wizard of Oz testing
- UX questionnaires, metrics and benchmarks
- Empathy maps, response cards

There are also techniques to co-create solutions together with selected users, making the feedback loop even faster. In a typical co-creation session, the team chooses one aspect of the solution, creates simple building blocks and lets a group of users work with them. This needs an example: To equip a police car, designers created cardboard models of the equipment to go into the car. They then took an old police car, their cardboard models and asked a couple of officers to place the equipment so it would be best. By doing this, they learned a lot about placement options, constraints, real world issues and elements of police life.

Things to be aware of:

Business analysts, market researchers, UX professionals all talk to the users and collect information but with a different focus. They reach different conclusions and there will be trouble if these do not align well.

Get really creative with how you create prototypes and test them.

Prototypes should make it possible to experience the future life with the product and they should be quick and cheap to build.

Try many solutions, your first idea will not always be the right one.

Evaluate the solution in comparison to relevant problems. Avoid letting users explore a prototype on their own or by using artificial tasks. Invest in creating realistic tasks and scenarios. Ideally, users should bring their own work and try to perform it with your prototype.

While doing this, you can expect the story to get clearer and more refined. You also learn a lot about the users, their lives, their work and their dreams, and get numerous ideas relating to the product concept.

Nail the concept

Once you really start developing the product, you will want to define a few fundamental cornerstones of the solution: information architecture, key elements of the visual design, error handling approaches and more. And you will want to test this with users.

Methods that help with meeting with users include

- Scenarios, visual scenarios
- Wireframes, lofi prototyping
- User/usability walkthroughs, hallway testing
- UX questionnaires and benchmarks
- Empathy maps, response cards
- Participatory design workshops, card sorting

Things to be aware of:

You cannot create a product concept without going into details. Take a few interesting key examples (key path scenarios) that require design and create the concept based on these.

Visual design is just like a nice sugar icing on a cake. It sells but a bad cake stays a bad cake. Thus, don't do pixel perfect. Use hand-drawn sketches and wireframes to elaborate the concept. You get a better cake faster.

Try the hallway to get quick initial feedback. Just put your draft sketches on the wall, grab anyone available and let them give you feedback.

The first solution is usually not the best solution. Try different approaches and involve others.

Again, realistic tasks or tasks that people bring with them are what you want to evaluate.

From these activities, you can expect to identify the architecture of the user interface. You will also learn a lot about users, about your story and you will discover numerous requirements and details of the product.

Work out the details and fine-tune your solution

Whenever you need to implement a user story, you will also need to work out the details regarding exactly what to do and how this piece fits into the overall product that has been developed so far. Welcome user input again here. Some methods that help when meeting users:

- Hifi prototyping, lofi prototyping
- Hallway testing, user/usability walkthrough

- Usability testing & usability lab
- Business walkthroughs, pilot installations
- Life A/B testing
- UX questionnaires and benchmarks
- Empathy maps, response cards

Things to be aware of:

When working out the details of a user story, start with the real needs and try to identify the simple solution. Avoid implementing wireframes that a UX professional created alone in his cubicle or requirements a BA wrote after listening to a stakeholder. Work as a team: what does the user want to achieve in this story and what is the best way of achieving this, and what should you therefore build.

Fine-tuning the product as a result of a story will also require changes being made to things that are already done and ready.

Let users use the system once a user story has been implemented. Provide a test system so you can measure parameters and obtain feedback.

Setup a simple usability lab if you need to optimise specific aspects. Users can work undisturbed here and video recordings allow you to analyse the details of the interaction.

Not every function has the same importance or difficulty. Test and optimise only what really needs to be optimised.

Don't let users be the first to test. Take a real life story and play it through yourself.

Who should meet with users?

Very simple answer: the team. Not the BA, not the RE, not the PO, not the UX person. You cannot delegate the responsibility for "Great Product" to one person. If the team does not have a sound common understanding of what is really important for users, those with little understanding will introduce small glitches and big blunders. Practical solutions need some compromises and UX skills are excellent assets in a team. But even if UX persons have the most exposure to users, do not let them have the sole contact. Always include other members of the team.

Options on how to get to users

Getting to users can be quite difficult. Here are a few ideas (including how not to meet users and still call it user-centric):

Simulate users, i.e. do it yourself and put on a user's shoes. Ok, not really users but at least you try to look through users' glasses.

Replace users with anybody over the hallway. Still not real users but at least people who are not experts on the system and will point out issues to which you have become blind.

Replace users with people who know the users: Trainers, retailers, support, service engineers and similar. Even though they are not users, they may have good insights about users.

Hold informal meetings and use your network to reach users. An easy way of getting access to users for a short session. A network is obviously needed.

Hitch-hike on field trips somebody else is doing (e.g. a service engineer, sales people)

Go to a place where you are likely to find users and get people to talk to you.

Go up the hierarchy to request users for a session.

Request users as members of your team for 30% to 50% of their time. Make sure that these users are still doing their normal job so they don't unlearn being users. Preferably experienced and expert users rather than novices.

Establish a user pool that you can use for sessions. Be aware that these users are usually locals and do not represent the whole world.

Use recruiting companies to organise "users" from their pool. Needs a precise profile describing what you require.

Open up your development to the general public and let users join the discussion.

Install your development team right next to your users, so you can just walk over and talk to them. There is no better way.

Need a final word?

Here it is, adapted from Google: Focus the team on meeting with users and everything else follows.

By Markus Flückiger

Some inconvenient truths about the digitalization of your business

Many companies today want to use digitalization to develop completely new – ideally disruptive – business models. In most cases, they intend to complement their product-related core business with "digital value-added services". In this article I summarize some inconvenient truths that I have come across time and again. Let yourself be demsytified, so that you can start the adventure of digitalization with realistic expectations. Let's start with a basic realization – the first of eight uncomfortable truths:

1. You are not disruptive with your digitalization strategy, but only part of a herd.

I know it sounds tough, but it's true: All managing directors today want the Internet of Things and Data Analytics. Everyone wants a platform and talks about Minimum Viable Products. All of them demand "fail fast!" from their teams and have conducted design-thinking training courses. Smaller companies create "Digital Business Innovation" jobs, while larger ones build up entire incubators and make the obligatory detour to Silicon Valley. That's fine. But, unlike a few years ago, such measures are already mainstream today. It is right and important to deal intensively with these topics. Initially, however, you will not achieve more than the competition. But if you face the uncomfortable truth, chances are good that you will be more successful than the herd.

When it comes to digitalization, I also perceive an over-enthusiasm to adopt new methods. But beware:

2. **There is no such thing as a silver bullet to realize radically new ideas.**

Unfortunately, one need is repeatedly expressed: "You told us about Design Thinking, Lean Startup and Scrum. We've heard all that before. Don't you have a new method we don't know yet?" Here I must loudly shout "Stop!". It is not important to use a method that no one else has. A structured approach is important and sensible, but the team composition, top management support and a positive failure culture are much more critical to success. Many managers long for new, life-saving methods because they feel the following uncomfortable truth:

3. **It is much more challenging to use agile and lean startup methods than it sounds.**

Agility and Lean Startup sound like casual and fun ways to work. This gives the impression that anyone who is sufficiently motivated and trained in a hip method can work this way. Unfortunately, these methods are anything but easy to implement and, firstly, require a great deal of discipline on the part of all parties involved, including the thorough formulation and validation of hypotheses. Secondly, fundamentally different leadership styles and a corresponding culture are needed.

Let´s take the popular business model canvas as another example: It is both a curse and a blessing of the canvas principle that every team can fill out a business model canvas in a 60-minute brainstorming session and go home with the good feeling of having done something cool. After all, you always have a result. I claim that 98% of business model canvases end here and are therefore no more than a nice finger exercise, because the laborious work is just beginning: deriving hypotheses, formulating tests, arranging appointments with users, validating hypotheses, iterating the canvas, etc. This takes a lot of time, ties up resources and is usually quickly lost in the urgent daily business of the team members.

One reason why the complex validation of the hypotheses is often skipped could be the following problem:

4. You do not know your customers as well as you think you do.

When a manufacturer wants to digitalize its business, it is almost always a question of developing digital value-added services around the product. If you want to offer your customers valuable (i.e. paid) services, you need to know both the small and the large problems your customers have every day in the context of using your product. However, this is a completely different type of knowledge from that which you previously needed for the development and distribution of your products. There are many treacherous "unknown unknowns" here, i.e. you don't even know what you don't know about your customers. And, until now, you didn't even need to know because things like multi-level sales or tendering standards around the "product view" were enough. Suddenly, however, the focus is on a service view, networking and new "customer touchpoints".

Suitable methods for this include Design Thinking for idea generation and the User-Centred Design Process for implementation. But what is much easier and what everyone should do is to just drive out for a day with the service technician. You will learn more about your customers than you can in all the training and study courses put together. That's a promise.

Understanding your customers' needs in the context of digitalization is a challenge for your entire organization, from development, product management and sales to service. But that is not all, because the following is unfortunately also true:

5. Digitalization does not fit into your existing organization chart.

Digitalization projects always bring with them the great challenge of being inherently cross-cutting. This makes the execution of digitalization projects very demanding, tough and time-consuming. At the same time, the "immune system of the company" prevents truly radical new ideas from emerging in the company – or it ensures that these ideas are increasingly watered down in the implementation process, so that in the end not much that is really new or even radical remains.

If you are aware of these tendencies in your company, then the next truth may even give you a sense of relief:

6. Disruptive business models of the calibre of Uber and Airbnb do not come from Europe.

Many industrial companies would like to have a big hit in their industry. No wonder, because in every management lecture the usual suspects from the consumer world are cited as models of disruption. But it is no coincidence that the biggest ideas come from Silicon Valley, because nowhere else in the world are so many innovation drivers concentrated: the deeply rooted and serious claim to improve the world (for example, Google only tackles problems that affect at least one billion people); the availability of venture capital and the willingness to invest enormous sums; the quest for "moonshot projects" (Elon Musk: "I want to die on Mars."); the openness to exchange ideas and accept failures; the density of top research institutions; the tolerant lifestyle in the most beautiful surroundings and much more.

Only rarely does a disruption worthy of the name come from another continent. Can you think of an example? The strengths in Germany and Switzerland, for example, lie elsewhere: in quality, thoroughness, safety and longevity – to name but a few.

The following truth may explain why the United States, with its pioneering and discovering tradition, is more successful than anyone else:

7. New business models are not designed, they are discovered.

Most companies expect a clearly defined path to success in the digitalization process. But there is no such thing. Look at how your company has come to the business models of today. You'll find in more than two-thirds of cases that the founders had originally started with something completely different and only gradually discovered and satisfied the needs of the customers which today make up their core business.

Typically, this observation also applies to digitalization projects: You start at one point with a (supposedly) good idea and then go out to the customer – to learn what he or she really needs. The goal from the outset must therefore be to discover the customer's real problems and realise that your company can make a significant contribution to their solution. This is the core of the "Lean Startup" approach.

The question arises as to how digitalization projects can nevertheless become a business case. It is often the case with monetization because:

8. Companies cannot grasp the business models of digitalization in their existing business logic.

Companies know that digitalization will fundamentally change their business. However, they still want to display the result in their conventional business logic. Crazy, isn't it? This is only possible with an incremental innovation in which there is experience from the past and assured expectations for the future. Then, for example, a ROI can be calculated quite reliably.

At the core of digitalization, however, are two completely different things: data and the customer interface. Both cannot easily be monetized via the industry's existing sales channels. The success factors are rather the development and use of multi-faceted markets, the realization of network effects as well as fast scaling and large reach. Digitalization creates the basis for new business models that are not yet foreseeable. The appetite comes while eating, so to speak, and the best ideas come from working together with third parties.

What do we do with these findings?

First: Don't digitalize your business. Instead build a new digital business. See the difference?

If you digitalize your business, you are a) cementing the status quo and b) most likely not making use of the full potential of working digital. Working digitally allows you to totally rethink your business, the products and services you offer, the way you win, interact with and keep customers and who you do business with. Here are some tips on what you need to increase your chances of success:

- The deep-rooted conviction among top management that digitalization makes new paths possible and necessary – and the willingness to go down these paths with courage.
- Interdisciplinary teams that are free to break new ground, with enough freedom and distance from day-to-day business, and to make mistakes in the process.
- The motivation to identify customer problems before thinking in terms of solutions. Usually it is not ideas that are lacking, but well-understood customer problems.
- The ability to learn openly and together with customers and partners what each party's individual path to digitalization looks like for their own added value.
- A positive error culture and the agility to quickly and iteratively move from the initial solution idea to a viable business model – before the project runs out of money.

By Moritz Gomm

Team fit

> *"The human race is filled with passion. And medicine, law, business, engineering, these are noble pursuits and necessary to sustain life. But poetry, beauty, romance, love, these are what we stay alive for."* John Keating (Robin Williams), Dead Poets Society

Science tends to measure and quantify things. By making them concrete, tangible, it sets itself boundaries. But passion breaks them. That's why they work so well together and why they can't "be" without each other. For engineers to perform in a manner that is beyond ordinary, an organization must provide a system that will foster the passion. That system is culture.

Culture is a group phenomenon. It manifests itself through shared values and behaviours and is experienced through the norms and expectations of a group. When in harmony with personal traits, it provides an environment that empowers people to perform at their best. A degree of that harmony is called *team fit*.

Software engineering is a highly collaborative process. From the time when work to create software is begun, in the discovery phase, until its retirement, we're in constant contact with various stakeholders. Furthermore, as we take on more responsibility, collaboration become more important.

Software is everywhere. Professional services take us to diverse industry sectors. Creating a solution that will meet the end-user's expectations, requires us to really understand the business, regardless of the role.

To work, contribute and progress in such systems, we need to be equipped with competencies that go beyond the technical. For that we need to look at the domain of emotional and social intelligence. The former provides us with skills to execute the task, while the latter enables us to take it through the process efficiently.

So, what is it, exactly, that we are looking for? This is a question I put to colleagues. After several individual and group discussions, we came to a consensus on a list of eight competencies, all of which are perceived as crucial in certain contexts:

- **Approachability:** a prerequisite for collaboration to happen is that you feel comfortable to approach your colleague no matter if you bring bad or good news.
- **Accountability:** *they may rely on you.* Once you have come to an agreement with your team mates, and the work is shared, they should be able to trust you to complete your tasks in the agreed time frame and quality.
- **Integrity:** may your actions be *at one* with your values, principles and beliefs, in every context. People of integrity always do the right thing. They are honest and feel comfortable admitting when they don't know something. They are open to different opinions and always welcome feedback.
- **Empathy:** only if you are honestly concerned about and understand the feelings and perspective of others, will you be able to establish a mutual bond. That bond is fundamental for establishing a feeling of psychological safety in the team – the foundation for every high-performance team.
- **Adaptability:** we live in a world of diversity. People we work with come from different cultures, have different opinions and professional/educational backgrounds. Projects we work on are from different industries run by various technologies. As an integral part of these processes, we absorb, analyse and assimilate a huge amount of information and emotions. We adapt.

- **Proactiveness:** in a time of innovations, products should not only be designed in response to impulses from the market in real-time, but one should also try to predict such impulses and act ahead of time. Engineers follow that rhythm and, assisted by proven practices, methodologies and the right technology choices, make this a reality.
- **Courage:** it takes courage to challenge the status quo and leave one's comfort zone. Those are the first two steps on the way forward – on the road between opportunity and success.
- **Endurance:** *good things take time.* Just try to remember how many things that made an impact, or of which you're proud, took a short time to happen.

You may ask yourself if it is realistic to expect someone to possess all these competencies. In my experience, it is. The difference being only which of these are closer to the surface.

By Marko Simić

The evangelist and the chameleon

Imagine this quite common situation: somebody, let's call her Andrea, works in a team that uses a particular tool that isn't Andrea's favourite. Andrea has a choice: she can either react according to her (probably unconscious) preference, or she can choose her reaction based on what she wants to achieve. Here is one way of looking at her possibilities:

Possibility 1: the whiner

Andrea will complain continuously about the tool she is forced to use. It is slow and a particular bug leads to undesired behaviour at least daily. And anyway, this tool is simply not as good as her favourite tool. Even after years, Andrea complains about the same problems. She has neither learnt how the tool works nor found any of the workarounds her teammates use nor taken any steps to replace the tool.

The whiner tries to convince others that her preferences are the best, but she doesn't take any steps to actually switch to her preferred tool.

The advantage of the whiner is short-lived: It is a good way to find out whether others dislike the same tool and (if you listen closely) whether there are any reasons why this tool might be the right one for the situation.

The disadvantages are that whiners are annoying and there's the risk that people stop listening to Andrea even when discussing unrelated topics.

Possibility 2: the evangelist

Whoever wants to listen (and everyone else, too) will hear continuously about all the advantages of Andrea's favourite tool. She does everything she can to have as many people as possible switch to her favourite tool, even if there are good reasons why, up to now, something else was in use.

The big challenge of evangelists is that they usually don't know the alternatives. This means that Andrea sounds more like "I don't want to learn about this tool" than like "I know something better". In the worst case, the functionality she points out as the biggest advantage has been implemented in an improved way in the very alternative she wants to replace.

The advantage of the evangelist is that many migrations need the tenacity of an evangelist. Without evangelists, we would still use the tools from the last century.

The disadvantage is that Andrea might overlook a far more important issue or a precondition for the migration.

Possibility 3: the tightrope walker

Where possible, Andrea uses her preferred tool. Whether she adheres to the team's conventions or not is a balancing act:

- if your environment differs too much from everybody else's, pair programming will become more difficult.
- most companies have rules about what can be installed/used. Because Andrea most likely doesn't know the reasoning for each of these rules, she has to consider carefully which rules she breaks.

- A tool you know well and that is configured to fit your preferences can increase your productivity and reduce your frustration.

The difference between a tightrope walker and an evangelist is that the tightrope walker focuses on her own productivity and comfort. She doesn't attempt to convince anybody, but simply uses whatever she prefers (as long as she believes that it is OK within the context).

The advantage of the tightrope walker is that her focus is not on a specific tool. This gives Andrea the freedom to look at the big picture and focus on the most important issues.

The disadvantage is that Andrea might miss an opportunity to improve the situation of the whole team.

Possibility 4: the chameleon

Even though Andrea has her preferred tools, she can work with many other tools. She adapts her way of working to fit the tool in use and continues to learn about each tool she uses. Given her experience with various tools, she can choose the tool that best fits the problem at hand, regardless of her personal preferences.

Like the whiner, the chameleon does not attempt to change anything. But the motivation is completely different: For the chameleon, the choice of the tool is far less important than many other things. The chameleon neither attempts to convince anybody nor takes any action to change the situation. This is not due to resignation, but because the chameleon adapts and is happy with many tools.

The advantages and disadvantages of the chameleon are the same as for the tightrope walker, with the added advantage that the chameleon has more experience with a broad range of tools.

It's your choice

Each possibility has strengths and weaknesses that may or may not fit your situation. So choose wisely: what's your intended impact? And which possibility matches your impact best?

No matter whether you work as a team member, as a consultant or in a management position: don't let your unconscious preferences rule your actions. Instead, adapt your behaviour to the requirements of the situation and continuously ask yourself: which of the possibilities above is the best to achieve my goals? Does my behaviour match my chosen possibility?

By Franziska Meyer

The evolution of support and operations team setups

Every software development project approaches go-live readiness at some point during the software development and enters the live phase. Actual end users start benefiting from implemented features and the sponsor gets the return on their investment. However, the phase before and after go-live can be intense. Aspects around the operational readiness, as well as organizational and procedural considerations regarding future support and operations are then at the heart of discussions.

There are many supporting factors that can lead to a project's success in this phase. One key good practice from our projects is to involve colleagues with experience in support and operations early on in the process. A service manager can provide valuable input during the bid phase in terms of release and incident management. And a DevOps Engineer can set up a CI/CD pipeline at the start. Such measures ensure a smooth, high-quality release to production.

However, when a software development project approaches the live phase, the team in charge undergoes a shift in focus and setup. In the following, we present a set of collaboration models for teams that are transitioning from initial software delivery to continued delivery with support and operations.

Fixed developer

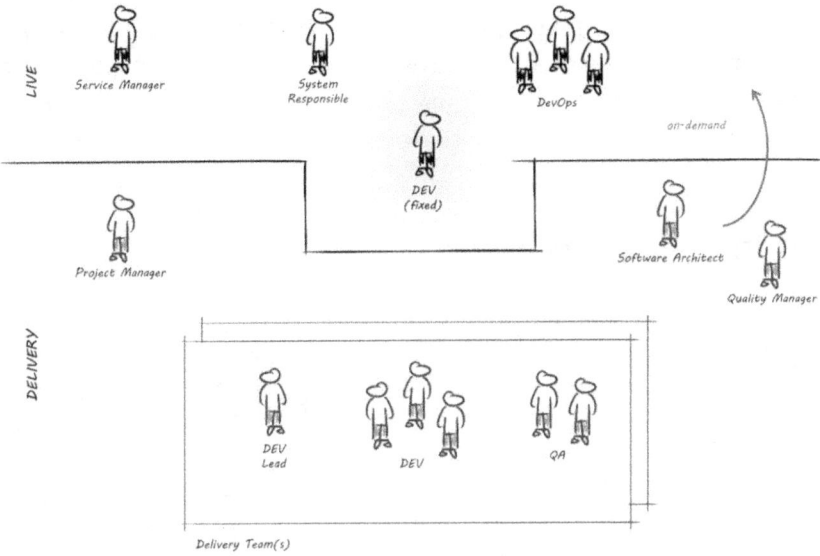

One or many developers may be exclusively assigned to support and operations activities. The benefit of this approach is that there is no capacity impact for ongoing development – as the fixed developer(s) are not participating in that. Also, it ensures that there is always a responsible team member available to respond to support and operations topics. However, this advantage comes along with a huge downside in terms of knowledge transfer. As the developer is not participating in ongoing development, supporting newly developed features in production can prove a challenge. Consequently, the fixed developer will, in any case, have to approach team members about issues arising in production. This effectively leads to a capacity impact, so the upside of this approach turns out to be questionable in practice. Furthermore, it is usually the case that a developer is not fully allocated to a project in this scenario. Because there are rarely enough incidents to keep an expert busy 24/7, a developer is assigned to multiple projects. This leads to cross-project planning conflicts, in particular when there are incidents to solve at the same time in all of them. Last but not least, this setup leads to an increased amount of context switching and will therefore reduce the overall efficiency of any person in charge. Given these practical implications, our conclusion is that such a scenario only works for projects where there is no ongoing development.

Rotating developer

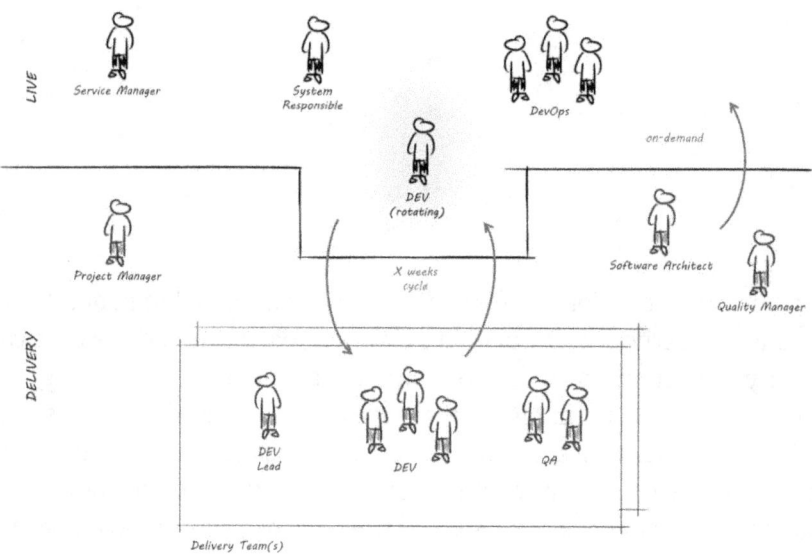

To mitigate the above-mentioned downsides, we experimented with a 'rotating developer' setup. Here, the responsibility for support and operations duties rotates throughout the team. The related advantages here are threefold:

- Between periods of taking their turn for support and operations duties, the developers are participating in ongoing development and therefore possess the knowledge to support incident resolution in production.
- Developers who fixed issues in production when it was their turn for support and operations duties will tackle the development of new features with a different perspective. The 'eating your own dog food' effect turns out to have a positive impact on quality after a couple of rotations.
- A developer can be allocated fully to one project, which mitigates the negative effects of the fixed developer approach. However, this collaboration model also has its downsides. While taking their turn for support and operations duties, developers cannot be fully assigned to backlog work and, due to the hardly plannable nature of production incidents, it can happen that even the reserved capacity is not sufficient. In addition, this model is a challenge in situations involving a broad spectrum of technologies. That is simply because one single developer is rarely an expert in all areas. But, given a less diverse technology mix, this model is well-suited to projects with strict delivery timelines – as the rotation model introduces appropriate planning means.

One Team

The motivation to continuously improve our team setups has led to a third model. It integrates the benefits of former models and attempts to mitigate two remaining downsides. As noted earlier, it is a challenge to take over responsibility for support and operations as a developer in projects where there is a diverse technology mix. Furthermore, despite having established a lean knowledge transfer due to rotations, there is still some overhead involved, e.g. on a handover. We therefore started to experiment with team setups where the whole team is continuously in charge of both ongoing development and support and operations activities – a model therefore referred to as the 'One Team' model. Once an incident comes in, the team decides who is best suited to tackle the issue in terms of knowledge and capacity. Therefore, this model is suited to projects with a diverse technology mix. Furthermore, it significantly increases efficiency because it eliminates the need for handovers and it benefits from having a team that is responsible for achieving a common goal. This team spirit releases supportive behaviours; for instance, even if a specific team member has taken over the task of resolving an incident in production, the remainder of the team will provide more input and support than in a scenario where this responsibility is delegated to a single person. Furthermore, the above-mentioned 'eating your own dog food' effect scales in that situation. A team that is very aware of the challenges of any given software solution in production will take very informed decisions regarding new features. However, this collaboration model also has its downsides. The number of issues to resolve in production impacts the velocity of a team and it can happen that a Sprint goal is jeopardized. Therefore, this setup is best suited to projects in which the priorities of delivery and maintenance can be balanced. If it is possible, at times, to postpone the delivery of some, less crucial, new features for the greater good of stable operations in production, then this model is the best choice. In fact, experience shows that no other model allows for a faster and higher quality resolution of incidents.

As you can see, the collaboration models have undergone an evolution over time, starting from the fixed developer model as a logical and simple setup to start with, progressing to the rotating developer model and finishing with the One Team model as a logical conclusion: The overall goal of each of them should be the same – sharing responsibility as one team.

By *Tijana Krstajic, Guido Angenendt*

The house of the six wise men

An old saga tells the story of the different levels that have to be taken into account when designing a new solution. It goes like this:

One day, an ambitious young shipbuilder came to the House of the Wise Men. The queen had personally sent him. He was to learn there how to make the most useful ships possible.

In the forecourt of the building, the shipbuilder met an old man sweeping up leaves. The old man showed him the way to the entrance, where he was already expected.

The shipbuilder entered the first chamber.

"Are you here to get some advice from me?", asked the first man quietly. "I can indeed give you some. The most important thing is how useful your ship will be for your mission. Think carefully about what you want your ship to be able to do. Does it need to have oars, in case there is no wind? Does it need to have weapons, so that it can defend itself? Have you given any thought to navigation? Many a ship has run aground because it unintentionally entered shallow waters. You have to weigh up what is especially important for you. If your ship cannot do enough things, you will not reach your destination. If you want too much, it will go down. Take note: The functionality determines the usefulness."

Deep in thought, the shipbuilder went to the second door, which opened as he approached.

"There is nothing more important than the reliability of your ship," said the second wise man. "Is the hull sturdy enough, but also agile and fast enough to reach your destination? Is the ship built so that it doesn't capsize when the waves become high? Do the masts remain firm in a storm? What do you do when the wind tears your sails? Don't forget: A ship can only be useful if it is reliable and free from defects."

The shipbuilder was still thinking as he entered the next chamber, where the third wise man was waiting for him.

"The most important aspect is your ship's handling. If your helmsman cannot control the ship, you will not be able to avoid obstacles. You have to be able to set the sails and also take them down again quickly, if the wind gets too strong. Otherwise you are doomed. My advice is: Make sure your crew can handle the ship and the equipment before you go to sea. Because, remember: All the technology in the world is only useful if people can use it."

The shipbuilder was excited in anticipation of the advice waiting for him behind the fourth door.

"Your priority has to be sustainability," said the fourth wise man. "Be careful regarding from where you take your timber, otherwise one day there will not be any good trees left. Will the tar pitch that that you need for your planks poison your village? That would result in nobody being left when you return. Does the yarn for your fabrics come from honest sources? Only what pays dividends in the long term can bring real benefits. You have the choice: You can be highly respected or ostracised."

This made an impression on the young shipbuilder, as he finally went through the fifth door.

"The greatest importance should be given to the impression your ship makes on people. Otherwise you will not be able to find a crew that is prepared to sail with you. Your ship has to radiate power, so that pirates do not consider it easy prey. Be sure to make it graceful. Remember that you will spend many days and nights on the ship. It should be a pleasure rather than a hardship to spend your time on the ship. Pay attention to every little detail, because beauty is an important asset. Everything that surrounds us must be appealing, because without beauty it is of no use."

When the young shipbuilder left the building, he again passed the old man sweeping up leaves.

"So," asked the old man. "Are you satisfied with the advice you have received?"

"I'm confused," replied the young shipbuilder. "Each of the five wise men gave me a plausible suggestion regarding what to focus on in order to ensure that my ship can be of the greatest possible use. So now I'm really not sure where to start."

"Could you give me a little bit of help to sweep the corner under this shrubbery? I'm old and can hardly reach there," the old man asked him.

The young shipbuilder took the broom and started sweeping. "These leaves get stuck. Your broom needs to have fewer, but harder bristles to be able to sweep here. And the nice handle looks appealing, but constantly slides out of my hand. Your broom is most likely much better suited to cleaning a doorstep. I now understand what the first wise man meant: It is the functionality that determines everything!"

The old man smiled. "Come back tomorrow, and we will continue with our work."

By Michael Richter

Time to say goodbye

Setting up a project in an efficient way is not an easy task. It involves sharing the vision of the project, and ensuring that everyone is aligned and has a common understanding of how the project will be managed. It is now seen as good practice to have a kick off workshop, not only for an hour but for a day, with teambuilding and vision refinements, not only once but depending on the project duration at regular intervals.

But how things look at the end of the project? Is everyone aware of how long they will be involved in the project? What are the tasks when the project is coming to an end? When is the last working day? What will happen with the project deliverables after project closing? Who is going to ensure operational startup, when and how?

The following often occur:

- Ops and Support became involved too late or not in an efficient way due to time constrains or other reasons. They have therefore not been enabled to take over tasks and responsibility.
- The transfer to operational mode is not proceeding because the project team is still receiving new requests and trying to handle these.
- People are happy to be contacted and needed, because there might be a certain kind of uncertainty in terms of how things are going and how much longer we will be involved in the project. Every change brings a certain kind of insecurity.
- Most often, knowledge transfer and handover do not take place, but resentment due to unclear responsibilities takes over.
- Sometimes the project manager simply says there is no more work and there is no information flow any longer. Project members feel dropped, not appreciated and a degree of frustration sinks in.

- Something similar might also happen if project members are assigned to a new project and expectations are not aligned between the project members, and the former and new project managers.
- And, last but not least, a completely different aspect: people get frustrated because of all the experiences gained from a project, as they realise that the new project is making exactly the same mistakes right at the beginning.

Based on these insights, I would like to give you 6 suggestions regarding what can be done to reduce or avoid the previously mentioned situations.

1. Get operations and support involved as soon as possible.
   With this I do not mean just invite them to the team meeting, but also ask them what their expectations about the project are and what they need in order to take over responsibility at the end of the project. Ask them for document reviews and involve them as testers. Create informal documentation of all issues and decisions taken as the basis for further reference and a handbook.

2. Give the transition phase a name, and define, plan and communicate this phase.
   Just like the development needing to be planned in advance, the transition also needs to be planned in advance. Official communication that this phase will now start and what the achievements of this phase will be help all the involved people and provide some kind of orientation.

3. Redefine roles, responsibilities and meetings in the team and communicate the results.
   With the new phase and activities, responsibilities will change. Maybe a business analyst will no longer be responsible for just

requirements but also for training, and the support provided by a support team or a test manager will no longer be needed. Some tasks will be taken on by operations and no longer performed by the project team itself. Therefore, a new definition of roles and responsibilities in the team is crucial, as well as communication of the new assignments. Team meetings or defect meetings may no longer be required. Therefore, meetings to discuss existing production issues, progress within training sessions or performance and usage statistics are more relevant. From my perspective, it is better to discontinue previous meeting series and setup new meetings with new agenda and participants instead of reusing old ones, as this ensures an official shift.

4. Establish contact with the team members and their superiors. Sometimes team members are 'lured away' to switch to another project. This is obviously not always the case but as a project manager I recommend discussing with the superiors by when the team member will leave the current project and to what extent the person will work in the new project or can take on new tasks outside the current project. This agreement should be made between the superior, project manager(s) and team member in order to ensure that they all have the same view of the situation.

5. No project without any learning elements.
Invite all the team members to a lessons learned meeting. Ensure that the meeting will either be split into separate ones if there are too many people involved for a common retro to be held, or get someone to help you conduct the lessons learned workshop. The workshop could be organised as a "world café". Good practice is to make the project stages visible by putting cards on the wall for each month and each event during the

project duration, making collages or writing a story/poem. As a source of information, you can use the monthly status reports or your project diary. You will be amazed about what things took place (and you about which you have already forgotten) and what you and the team achieved. One important point for me is that at the end of the meeting every single participant has the possibility to write down his/her own lessons learned and take these notes with them into the next project. Maybe you can prepare some cards beforehand and hand them out as a template that they can fill in. You can also ask your team members either for feedback about your performance or if they want you to provide them with feedback about themselves. In order to ensure you are prepared for these activities, it is helpful to write a project diary and note some keywords each week regarding what happened, what was good, what was bad, what was learned and also, using your private project diary, what happened to all the team members, so that you can see their development along the way.

6. "Goodbye and thank you for the fish".
   Project managers often forget to allocate some budget for celebrating project success or having a small give away as a "Thank you" to the project members. Regardless of whether or not you have any budget for this, it is a good habit to invite your team members in advance to a project closing meeting. The purpose of this meeting is to officially hand over all responsibilities to the operations team and relieve the project team of their responsibilities. That is also opportunity to reflect on the past months and perform a project review. Summarise the findings from your lessons learned meeting and share the impressions of that. This is the moment to say thank you. Maybe with some personal words for each member, a card with some keywords about what was special about the person during the project

duration, photos, etc... There are many possibilities to show recognition without needing expensive gifts or events.

7. As a conclusion, in my opinion a successful project closure must be prepared already at the beginning of the development phase. Documents should be updated with each sprint or iteration and each learning and insight should be documented in a way that allows it to be handed over to support and operations to enable them to solve issues by themselves. Writing some kind of project diary for the whole project might also help at the end of the project to remember what happened, what was done to solve issues and if these were successful. Just as the project has a planning phase at the beginning, it should also have a transition phase when it is coming to an end. Here I dare to make the statement that transparency provides certainty. Talk about what's going on and what the plans are, and involve all the team members and stakeholders, as you (hopefully) did at the beginning of the project.

By Sabrina Lange

Transitioning systems engineering into the lean-agile world

Lean-agile has swept over and transformed software engineering during the last decade. Despite additional hurdles when implementing lean-agile in the systems and product engineering field, the transformational force and the benefits are comparable. Many companies have started to adopt agile practices outside the software engineering discipline. Most of these are at the beginning or in the midst of the transition. Zühlke has provided consulting services for a leading global manufacturer of laboratory equipment and consumables, with regard to transforming the business units and 10 development competence centres worldwide to allow the advancing of sustainable profitable growth by implementing a lean-agile innovation process. The following article summarises experiences and lessons learned from this agile transition project.

At the top management level, agile is a means not a goal

The customer has been very successful in the past two decades in defining quality standards in the industry and growing healthily by pursuing geographic expansion. As is often the case, success brings complacency and in this case insights into the real needs of different customers and the predictability of project outcomes and timelines needed improving. Just making the product 10% more accurate or faster no longer created sufficient customer value. The technical limits of improving product performance had been reached. Breaking through these made the products excessively expensive and the development effort unpredictable. Agile, with its focus on customer value, early customer feedback cycles and delivering working prototypes with a short cadence promised to be a suitable remedy to many of the weaknesses in the organisation. Launching products that excite customers and users in a regular and predictable tempo was required to secure the future growth and success of the company. When transitioning to agile, we always have to steer the transformation based on such overall goals of our customer.

To start, organisation beats process: What's in it for me?

The challenges that were met were as diverse as the scope and size of the various competence centres. Writing software applications, creating complex multi-disciplinary systems and designing plastic components to a precision of a few thousandths of a millimetre are examples of the range of development work performed at each plant. Some R&D competence centres have around a dozen engineers while others employ more than 60 people. At all plants, the R&D teams constitute the technical knowledge centres of the plant and besides developing new products they fight obsolescence in current products, evaluate and commission new production equipment, support production, suppliers and sales and much more. Agile poses different challenges in such a diverse environment. It needs to be carefully worked out what agile means for each team, how it can be implemented and what the path from the current situation to the agile world looks like. Only when each person involved knows their role in the agile process, is he/she ready to discuss the details of the agile procedure and the methods.

Stepwise adaptation of agile thinking and methods

As the situation regarding customer insights suggests, few of the pilot projects introducing agile at the customer started with an established set of requirements. They started in an explorative agile mode, where developing the requirements, rating customer value and defining a suitable product concept constituted the major scope. However, to start the agile cadence with its cycle of sprints, an initial product backlog was needed. From that, each team worked its own way down to the team/sprint backlog that could drive the first sprints. Now the agile sequence of ceremonies was introduced, and the duties and behaviour of the roles became more refined. After those basics had been adopted, more elaborate practices, such as project increment planning, were introduced and trained. Once the product positioning and conception was finalised, the mode of the project changed to an agile execution, with further refining of requirements and design being conducted in parallel.

Simplicity, adaptability and acceptance are key

Agile requirements engineering proved to be particularly challenging in the diverse environment at the customer. Generally, the products and systems there exhibit more and wider reaching interdependencies than many software applications. Each engineering discipline had different perceptions regarding how a set of requirements should look like, and which the engineers had to work on during a sprint. Concepts from the software engineering world like features, enabler and stories were not always easy to transfer into the worlds of mechanics and automation. The separation of functional and non-functional requirements provided little additional information or help in designing physical parts. We therefore ended up with two levels of requirements that every team could translate into its own world and work with for planning and realisation:

Product Backlog Item: Defines the characteristics of the product for customers, users, production, logistics and service; owned by the Product Owner and agreed with all relevant stakeholders before implementation; a complete set represents the requirements specification (Lastenheft) of the product.

Team/Sprint Backlog Item: Split and detailed requirements and specifications ready for implementation during a sprint; owned by the Subsystem Owner and agreed with all team members; basis for sprint planning and execution; a complete set represents the technical specifications (Pflichtenheft) of the subsystem.

That concept proved to be sufficiently agile to support the agile cadence and adaptable to the needs of e.g. plastics engineering, which necessitates a complete set of requirements before finalising the design and releasing the tool making.

Collaborative product strategy development is a mixed blessing

Many agile pilot projects were started without an agreed product positioning and vision. Although many engineering and application team members considered it interesting to be involved in such strategic clarification, the ability to contribute differed widely and not everybody could cope with the degree of uncertainty involved. However, management accepted the low efficiency during the exploration period and innovative approaches and product concepts regularly resulted from the collective struggle to find better solutions.

Estimation, velocity and definition of done remain difficult and critical for efficiency

Besides the boost in motivation that results from the involvement and ownership of team members, the undisturbed estimation and execution of development tasks in a short cadence during a sprint is a main driver for increased performance in agile development. However, gauging team velocity, nailing down estimates and committing to the completion of deliverables remained challenging leadership exercises that needed constant attention in a cordial company culture.

Who is part of the development team? Who works in cadence?

To focus on project development work, we stipulated that each development team member should work at least three full days per week on the project. That was a daring target for an organisation used to working on many projects in parallel. The support of the top management for this helped to make continuous improvements on this subject and to bring distraction down to a bearable, although constantly disputed level. The handling of interfaces to other functions like production, procurement and marketing was also challenging. In traditional project management, those functions are involved as extended team members and their tasks are planned and tracked on a Gantt chart. Fully including the representatives of those functions in the agile cadence of the development team is inefficient. Often, those functions are intensively involved only during a specific period of the project. We had to establish the awareness that organisational interfaces need to be explicitly managed and that there are shades between being "fully inside" and "fully outside" the project. Some organisational interfaces e.g. launch preparation, can run in parallel with the agile cadence of the development team, with their own (launch) team and coordinated by the project increment planning and Scrum-of-Scrum mechanisms. Tool making could be integrated into the development team during the respective sprints. Some interfaces need to be managed specifically by the responsible person. We continue to work on establishing practical guidelines for the differentiated handling of interfaces that might be less agile and rely on clearly specified inputs and outputs.

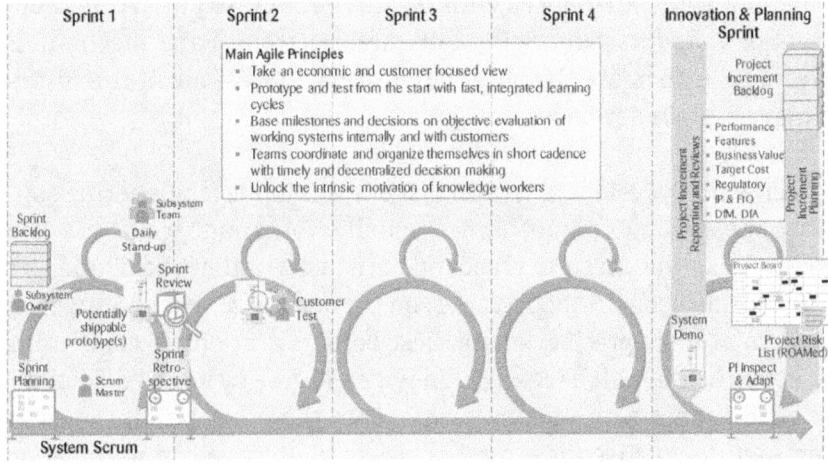

By Rolf P. Maisch

We are all engineers but work quite differently: software engineers, electronics engineers, mechanics engineers

What are your experiences with regard to interdisciplinary development of a product comprising software, electronics and mechanics? I would like to share the observations I have made about interdisciplinary system development!

It seems that developing within just one discipline (engineering software or electronics or mechanics) usually works quite well. But when these disciplines need to create one product together the challenges arise at the discipline boundaries: on the one hand, these boundaries (and, to be more precise, where these borders are defined concerning which function will be realized in which degree by which discipline) offer the opportunity for innovation! On the other hand, the most common root cause for project issues is ineffective communication at these borders!

Interdisciplinary projects require outstanding attention at discipline interfaces: primarily with regard to human communication!

But being aware of these boundaries is not enough. The working mode used by each discipline is just very different because the constrictions are very different. Some examples for a device development comprising software, electronics, mechanics are (see figure 1):

Mechanical components and assemblies usually start with the final form factor and will be detailed subsequently. Going from the concept phase (alpha) to series development (beta) usually means a change of manufacturing technology (e.g. from additive manufacturing to plastics injection moulding).

Electronic circuits are usually realized in or close to series technology even in concept phase (alpha). The final form factor is usually considered only at series development (beta).

Software usually provides end-to-end functionality at the concept phase (alpha), maybe even with mock ups. These existing features are extended and stabilized during the series development phase (beta).

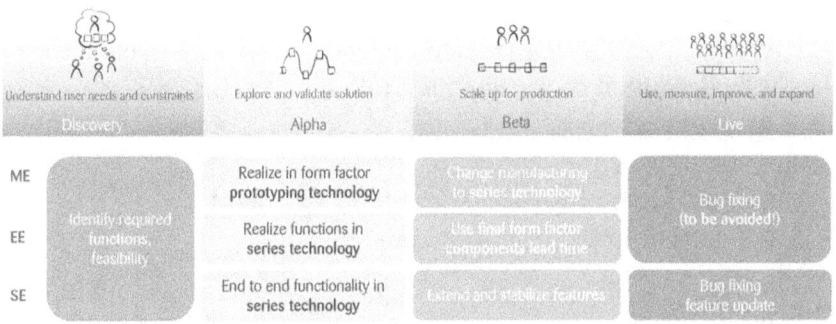

Figure 1: Comparison of main drivers per discipline during product development lifecycle

The next interesting cause of communication pitfalls are ambiguous terms! Are you sure you are talking the same language as your colleagues? Check out some terms that illustrate this problem (the term "HW" refers here to electrical and mechanical engineering):

"Iteration"
- HW: related to maturity, not cadence
- SW: cadence or repetition
- Proposal: use "sprint" instead of "iteration"

"Prototype"
- HW: typically a sample for verification (close to launch date of product)
- SW: e.g. UX-prototype, proof of concept, throwaway prototype (at the very beginning of development)
- Proposal: use a specific prototype like "verification prototype", "UX-prototype", "throwaway prototype"

"Design"
- HW, SW: aesthetic design
- HW, SW: construction, plan for building

"Increment"
- HW: typically an HW "increment" refers to the next level of maturity
- SW: by definition each increment takes one sprint

"Subsystem"
- HW: logical meaning (like description of structure)
- SW: logical or implementation unit

"Integration"
- Does this mean in the functional (HW, SW) or spatial (HW) sense?
- At component (HW, SW) or system level (HW, SW)?

Talking about communication also requires thinking about the different types of meetings and an effective culture: pay attention to whether the right people talk about the right topics, e.g.

Standup
- It is your responsibility to ensure that everyone attending understands the (high-level) information you are providing, including the limitations of your knowledge.

- Set the stage for your information: give a context (one or two sentences)
- Provide specific information about the context
- Explain possible consequences
- State what you are planning to do concerning this topic
- For explaining a discipline-specific term, it is helpful to have a term of the week slot after the standup. Make sure to also document the term in the project glossary.

Technical sync
- Use a regular technical sync, e.g. weekly, for planning the big picture and details of the integration strategy. Discuss topics affected by more than one discipline, make the unknown visible!
- Document and communicate your integration plan (e.g. by using the maturity table described below)
- Include in this planning also the use of rapid prototyping means (EE, ME) for early integrations and physical samples. These samples provide insights and value!
- Use a continuous integration environment for the system to ensure robust artefacts where the new unit is to be integrated
- Reconsider what has been learnt and the assumptions that have been made when making decisions
- Further opportunities to avoid conflicts during the project lifetime exist in identifying internal requirements such as
- Interfaces to bring up
- Development-driven interdisciplinary tasks
- Consideration of how interdisciplinary support for bug fixing might work

There is a central communication tool that addresses the needs mentioned above: we call it "maturity table". This tool is a simple, but very effective, table describing the integration strategy with its steps and associated samples. It develops during the project lifetime and needs to be updated regularly. Typically, each insight to be generated (or realized device sample) is described by one column, each subsystem in one row with its required functionality and maturity (figure 2).

Integration Step	MotorMoves	SensorWorks	MotorWIP	Integration Steps	Mainboard	UserScenario1
Scenario(s) covered by intergration step	Motor moves reliable and precisely	Sensing concept	Motor integrated with Sensor System evaluation board	Scenarios to be provided by integration step	Actuating and Sensing w/ dedicated mainboard	scenario
Model setup	Motor+gear	Sensors+elect.	Motor+evalbrd	+Sensors	Eval→Mainbrd	IMo_3
Release date	DD/MM/YY	DD/MM/YY	DD/MM/YY	DD/MM/YY	DD/MM/YY	DD/MM/YY
Test	Check oscillogram	Check oscillogram	Check oscillogram	Check sniffed data	Check regression.	Test scenario 1, 2
Actuate piston	TFB...		TFB...		TFB...	
Sense temp.		TF2...		TF2...	TF2...	
Sense force		TF5...		Detailed description of degree of maturity and acceptance criteria per integration step and subsystem		TF6...
Orch. workflow						
Handle UI					TF...	TF7...
Equipment & deliveries	OTS controller oscilloscope	Oscilloscope	Oscilloscope	Oscilloscope, sniffer		Delivery by sponsor.

Figure 2: Maturity table as a central communication tool describing each integration sample (simplified example)

This table is very valuable for discussions between the project team and the sponsor as well. Based on functions and subsystems, it allows you to trigger the necessary discussions to avoid surprises during device development. My three most important takeaways for you are:

The more disciplines are affected by a system, the more sprints are necessary to generate a common understanding of the system and the neighbouring discipline. Usually this is mission-critical.

Use a systematic approach to ensure the right people are talking about the right topics (e.g. an adapted meeting culture and the maturity table).

Don't assume anything, especially about other disciplines – just talk!

By Thomas Weber

What's wrong with: "I don't write any tests, since I am not a tester"?

"Not a tester, so what are you then?" you might ask.
Causing offence in this way is generally not helpful.
Unless you are trying to attract attention, which is what I am doing in this article 😉

Let's digest the situation in detail.

A friend of mine attended my Scrum Developer class and became very enthusiastic during the "Testing" module where we talked about code quality, testing, test-first approaches, TDD and more. Boom! After that class he was on a mission to convince everyone that TDD is the only way to do things.

The first day back at work he talked about improving the team and trying TDD, and was on the receiving end of the following statement from his colleague: "I don't write any tests, since I am not a tester".

I know he handled the situation quite well, but he asked me for advice.

One thing to consider is the underlying question to this, which might be: "How do we get people to change their behaviour?" So here are my thoughts.

Consider your own conduct first

First of all, think about how you deal with things yourself:
- Why is the practice or tool that you are suggesting any better than the current way of doing things?

- Can you explain the value of the proposed change?
- Can you lead by example?
- Do you have enough patience and skills to teach others? I would try to work on yourself before trying to change others.

Roles?

I see that a lot of people are focused on their own role, forgetting the bigger picture of the team and the purpose of the work they are doing. I would ask these questions:

- Are we one team that focuses on the Sprint Goal? In a Scrum context there is no "tester", "programmer" or "architect", we are all professional engineers who deliver value through collaboration.
- No matter what, do we stand together and support each other?
- Are we doing whatever is necessary to deliver a usable product every Sprint?

Done?

Colleague: "I don't write any tests, since I am not a tester".
Ask: "How do you know when you are done?"

What is on our Definition of Done? How can we build a usable, tested and fully integrated product increment every Sprint? Are we doing that already? Why not?

Fast, automated feedback

Tests send you a message. They send you a message now and in five years from now. They tell you:

Is the code working as it is expected to or not?

This fast feedback is very valuable if you are working on your product, whether changing it, fixing a bug or adding a new feature. Think of these tests as development support. They guide your development efforts and make sure your development doesn't derail, allowing you to go faster. And the additional benefit you get in the future is an answer to the question: "Did we break something?"

Tests are an important kind of documentation

Documentation is needed, and one good way to document how software must work are tests. I emphasize must, since written documentation only documents how the software might work.

We have too often learnt that documentation can easily get out of sync.

Quality

Quality is everyone's responsibility in a Scrum Team. There is no QA team in a Scrum context, which means the whole Scrum Team is responsible for delivering high-quality software that works and is fully tested.

Quality attributes that are important:

- Does it work at all?
  Huh!
- Does it work well?
- Is it deployed and usable?
  Are the users able to access it?
- Is it useful?
- Is it successful?
- Does it make the impact we wanted to achieve?
  -> Yes! Value is key

Tests are code

Are you a coder?? Yes?

Tests are code. Don't wait for the "test automator" to test your code. It's more efficient if you write tests that drive your production code and test your work. With those tests you get the benefit of fast feedback and your code gets tested and checked on every push. Additionally, over time, you will know that you have not broken anything and you will be able to sleep better at night.

Still not willing to write tests?

Ask:
- How can You help?
- What can You do?

You can always get them coffee.
Show support. We are in this together.

By Peter Gfader

When machine learning meets software engineering

Software engineering (SWE) and machine learning (ML) have recently become neighbours in academics as well as in professional services. They are so closely adjacent, indeed, that some authors dare to conclude that SWE and ML are simply different ways to achieve the same result, at least within the boundaries of particularly well-defined problems such as rule-centred functional problems, for example. The thinking goes like this:

Software developers encode domain knowledge into explicit, executable rules, such as if-then statements. ML practitioners, on the other hand, take a sufficient number of input data examples and attach the intended results as so-called labels. Hence: labelling is the new programming.

Software developers compile their software artefacts to create executable binaries. ML practitioners automatically adjust the thousands or even millions of parameters of a chosen standard algorithm until that algorithm returns the intended output for their labelled input data. This so-called training procedure is the new compilation.

Software developers (hopefully) write a lot of tests to prove correctness and provide a stability harness for their code. ML practitioners closely observe key statistical properties of their input and output data.

It should be obvious from the above, though, that we intentionally ignore the not so subtle differences that are still relevant if you want to understand the whole picture. For example, machine learning cannot (yet) be used to devise or even create attractive and effective user interfaces; and rule-based systems, such as those to transcribe human speech for example, can no longer catch up with the performance of modern neural networks. So, there are differences. There is room for both professional practices.

And to meet the modern consumer's ever-growing appetite for smart applications, practitioners of both fields need to unite. Unfortunately, this sometimes turns out to be not so easy as the naive observer might expect. Indeed, if you think back, the difficulties should not even come as a surprise. If you are a seasoned developer, you may remember hearing sentences like "I'm a software developer, not a database admin" or "I'm a software developer, not a system operator". If you are new to the field, though, you may not have heard any such thing, and the reason is probably that there is already a cure for this problem. It's called DevOps and it appears to be a very satisfactory common field for all participants; that's probably why they met there in peace. DevOps has effectively managed to unite software engineering and the classical related fields.

Nowadays you may hear "I'm a data scientist, not a software engineer". While that may be perfectly accurate, the need to emphasize the fact unfortunately implies the claim: "I don't need to care what it takes to use my results in production". On the other side, typical enterprise developers faced with the need to understand, write or simply integrate Python code find it hard to overcome their acquired resentment towards scripting languages.

The point is that data science, being exploratory by nature, requires an extremely expressive, interpreted language to effectively deal with the underlying uncertainty. Python has become the most popular language choice precisely because of its unrivalled expressiveness. Since the problem space is mostly of a functional nature, object-oriented design is rarely applied. Enterprise software development, on the other hand, typically deals with critical applications embodying deterministic complexity. And developers need to consider the fact that their systems need to be changed frequently, while a high-quality level must be maintained at all times. That almost certainly mandates a strongly typed, compiled language and some form of object-oriented design.

So, when we now hear people saying things like "I'm a data scientist, not a software engineer" (imagine the associated disgusted facial expression...), we can either wait for someone to come up with another buzzword like DevLearn or MLDev – or rather simply remind ourselves that we're all on the same mission: to deliver ever better, smarter solutions to our clients and their respective customers. By simply embracing diversity in both technical choices and practices and saving a good lump of openness and curiosity for our professional neighbours, we software engineers and machine learning practitioners can meet as friends and prosper!

By Wolfgang Giersche

Why every project should have gardeners

Just as a gardener takes care of his plants to ensure some fruit, we should also have people in the projects who help us to learn and develop so that we can move to the next level. Unfortunately, this is not yet common practice, so I would like to show how this could work in our projects, by staying with the image of a gardener.

Everyone started small. Even the tallest and strongest tree started as a seed and over time, thanks to a lot of sunlight, rainfall and good soil, it became a deep-rooted, reliable and sturdy tree. Very much the same also applies to us, people who are involved in project work, when we try to do our best to make the project a success.

We work in projects as subject matter experts, as experienced project managers or as newcomers that will need support from the team to grow to the next level. Roles and tasks are assigned based on our experience. Experiences from former projects are our soil. We are able to grow with new and demanding tasks, but may fail if tasks are overwhelming. If our specific responsibility lies within our comfort zone and things are just business as usual, we will not reach our limits. We need challenges, such as finding new solutions and new approaches, in order to grow. Just as a plant needs more and more space to become bigger, we need space and the opportunity to make our own decisions and take responsibility. We need staffing managers who see the potential of the project and also the required skills when staffing such a project with team members. Managers that are aware of the strengths and weaknesses of each person and if there is a realistic possibility that they can perform well and do a great job in the project.

In addition to soil, a plant also needs some kind of physical support right at the beginning. While the roots are still not deeply anchored, and the stem is still fragile, it is important that the plant is supported. It is exactly like that with us when we grow in our projects. When we have all these challenges and opportunities to grow, sometimes we make a mistake. Sometimes we do not fulfil the expectation of our stakeholders, do not communicate well, do not take the right decisions, and do not act as expected. These are the moments when we need support and orientation, when this situation takes us to our limits. We need someone who takes care of us, helps ensure we do not break in the storm and instead become stronger and are prepared for the next time we face a similar situation.

Most often, we expect the project manager to support and stand up for us, but this is not always the case. This may be because of personal reasons or political issues such as internal – external employees. From my perspective, it doesn't matter which role supports me, as long as I know right from the beginning that I can count on a person who is going to support me. This supporting person should be one of the project stakeholders or at least in the area affected by my activities. It could be the PL, the sponsor, a team lead, someone from operations, business representatives and so on. The crucial aspect is that this person is aware of the project and what's going on, that I am linked and in close contact with the person, that I have open and transparent communication with the person and that the person is a supporter of mine. To this end, bilateral meetings can be arranged or feedback requested.

Even if our soil is nutritious and we have support, we still need water, heat and sunlight to grow. Only with this can we become what we would like to be. Within a project, this means that we have someone close to us ("the gardener") who ensures that we deliver on time and with the expected quality, and that we not only choose the straightforward way but also try new solutions if the time and place are right.

Looking at the easiest way and what occurs most often in projects, people ask for help or reviews, or carry out pair programming e.g. consult peer groups outside a project to share experiences and ask for advice. As this is most often related to objective questions or tasks, my experience is that people are often happy to connect to others to improve the results of their work. In self-organising teams, team members often transfer tasks between them when required, in order to achieve the best match of skills and performance to the respective elements. Maybe you have also the possibility to be supported by a consulting coach or mentor. As an example, we had a setup in which the junior project lead was allocated dedicated tasks and areas for which he was responsible. Once a week, we checked the status and deliverables, and discussed possible scenarios.

The last aspect I would like to share may be the most critical one. All of you who have a garden with trees or roses will be aware of the fact that sometimes you have to cut back old branches to enable the plant to grow even more. Even if it seems as if the plant is destroyed and will never come to life again, by the latest in the spring it will be back again, bigger and brighter than ever before. And the same applies to us human beings. Sometimes we need someone to tell us we are doing the wrong thing, following old bad habits or should try new approaches.

We may not be pleased to hear that we are not performing well, but sometimes we have to decide to accept feedback and decide whether to follow the advice or not. All feedback should at least lead us to think about ourselves, even if we decide that what we did was actually ok, and we do not need to change anything. Possible ways to grow include, for example, asking for 360-degree feedback from team members, peers and superiors. Additionally, it is of great importance to always reflect on ourselves and on the feedback we receive. Therefore, it is important that we know our own strengths and weakness and find ways to further strengthen the strengths and to overcome our weaknesses. A personal mentor or coach can be very helpful in this regard; also, soft skills training or professional literature can help expand your personal horizons.

The conclusion of my recommendation is to check the staffing of the project team right at the beginning in order to find someone to take on this mentoring role. Not only the availability of the resources but also if they fit for the project. Sometimes it seems to be a good idea to have a "training on the job" setup. In this case, it is important to ensure that there is someone within the project who has the trainer role and to take this approach into account in the project planning (time and budget). At the same time, it is also the responsibility of each and every one of us to take the opportunity to grow by looking for our personal gardener or a growth concept as well as being prepared to be the gardener for someone else, and thereby provide support in the growth process.

By Sabrina Lange

Why you should create a paper prototype – and how to test it with your users

At Zühlke, we strive to work in a lean and agile way. We embrace the uncertainty by starting with a focus on exploring and learning. In interaction design, this can be achieved by creating, testing and iteratively refining a paper prototype.

Why test a paper prototype?

Some of the reasons are:

1. Getting the interaction design right should be your first priority: navigation, workflow, organisation of content and the terms used to denote domain elements. Simple black-and-white wireframes help you focus on these issues.

2. Action sequences or *scenarios* are paramount in interaction design. Creating a paper prototype makes you think about these sequences. Will the user understand how to perform the next step? And if so, how can they navigate back?

3. A paper prototype helps you get the most relevant feedback as early as possible. Do the users understand the navigation? Did you get their workflow right? Guess what: sometimes you won't have, but all you have to do to improve your design is change some wireframes.

4. Starting with a low-fidelity prototype means there's a natural focus on the big picture. Once the interaction basics have been validated, there will be time to fine-tune all the details of your design.

5. Creating a paper prototype allows you to involve your stakeholders in the interaction design and evaluation process. Schedule design sessions with them. Have your stakeholders observe the user tests. You can even change the prototype on the fly while you are testing it!

Creating a paper prototype

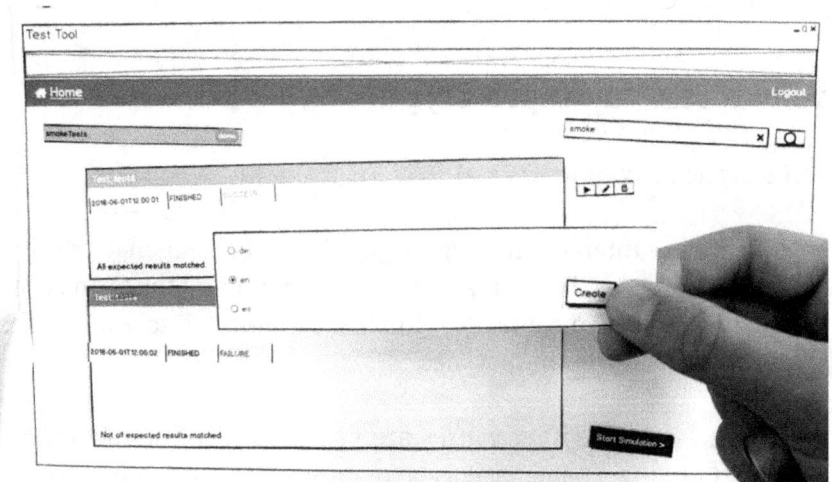

Here's the basic workflow:

1. Select the scenarios that you would like to test.
2. Create a set of wireframes, either directly on paper or using a tool like Balsamiq. Make sure to include all intermediate steps in your wireframes that a user would go through when performing the scenarios.
3. If one step involves adding a dialog on top of an existing page, create a wireframe "snippet" that just shows the dialog. This also works for other small changes on a screen.

4. The result is a stack of sketched or printed screens and a collection of snippets.

Conducting the test

Let real users of your system test the paper prototype. For a typical early-stage paper prototyping test, five representative users are enough to find the most important issues in your design.

Your goal during the test is to observe users performing the scenarios. It's very much like a regular usability test where you replace the computer with a stack of paper. To prepare the test, write an instruction sheet that asks the users to perform the tasks underlying the scenarios you used for creating the wireframes.

Ideally, one person "plays the computer" and prompts the participant, while another team member takes notes. Explain to the participant that you're testing the design, not them. Each time they get stuck or fail to understand something, you uncover a problem in the interaction design that needs to be fixed. Tell the participant to read the tasks and to perform them in the specified order. Ask them to "use" the wireframes as if they were a real interactive user interface. Finally, ask the participant to "think aloud", so you can follow their thinking process.

Never explain how to use the prototype. Resist the urge to help the participant if they're stuck. Use this opportunity to explore their thinking by asking "What did you expect?" or similar open questions. Only if the participant has been unable to proceed for at least a minute or so is it OK to point out how the task was meant to be performed in the prototype.

Take notes of everything that happens during the test. Protip: Write each observation on a post-it note. After the test, ask the participant to share any observations, opinions or suggestions and write a note for each of them. It's OK to discuss solution ideas at this point.

Evaluating the test

In the team, debrief each session directly after it's finished. What were the three main points to take away from it?

When all the sessions are done, cluster and prioritise your observations. Don't be too formal about this – the most important issues in your prototype will be obvious by now. Each cluster of observations needs to be addressed in your interaction design. So now go ahead and create the next iteration of your paper prototype. Rinse. Repeat.

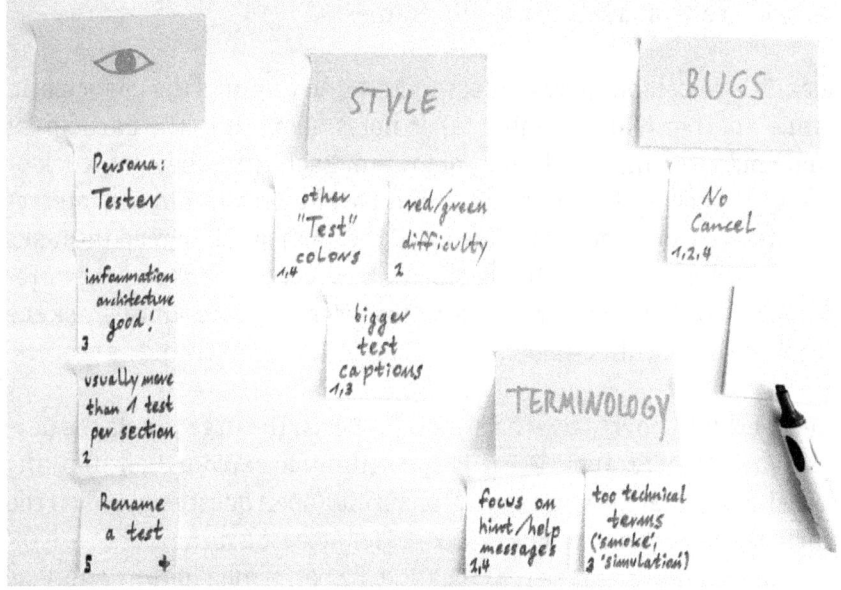

By Eric Fehse, Manuel Jung

You DiD what?

Is it outsourcing?! Is it team sourcing?! No! It's distributed development!

Dive into the topic of distributed development and get all the information needed to kick-start your project in a distributed development setup.

1. What is it?

Distributed development is a powerful way to carry out project research, development, and realisation across two or more physical locations. The difference between outsourcing and distributed development is that, in a DD setup, all the organisation is working together on the same level with a common goal of realising the project.

2. How does it work?

In the DD setup, there are multiple team members distributed to at least two locations. It's usual for the whole team to meet at one location where they have an opportunity to meet each other in person, meet their clients, stakeholders and get direct contact with all the key members of the project.

Transparency in the team is one of the most important aspects and the key element for creating a one-team feel. All team members should share ideas, experience, information, resources, and decisions.

Team collaboration is internet-based, which means that it uses online tools for daily work such as Skype, email, Jira, Slack, Whiteboard, Retrium, Trello etc.

Organisational tools are remote and co-located pair programming and knowledge sharing sessions. They help to make the team's knowledge base stronger and boost up their confidence and mutual trust to perform as one team.

3. What are the main benefits?

Scalability of the team is close to limitless since it is not dependent on one location. Finding skilled team members as the best expert for the job is easier in DD setup since it creates access to a larger pool of employees. In addition, it helps to keep in close proximity to the market and the customer. Cost benefits are positively affected since it allows a project to hire team members from other locations with lower rates but keeping the same quality of and desired expertise level. DD, with its distributed locations, increases competitiveness on the global market and allows companies to meet growing market interest. Companies have an opportunity to meet new people, travel and exercise knowledge sharing throughout the whole organisation. It also decouples employees and their work from their physical location which means no location boundaries. They can work from home or from any office (across the whole organisation). As the DD setup is the vendor's internal organisation process, there are no overheads for the customer.

4. What are the challenges and how to overcome them?

Language barrier – bilingual
English, as a common language of communication, is one of the best practices for distributed development setups. But as they cover the global market, the most common challenge is the language barrier. This comes mainly from the customer side, which is understandable and realistic. Simply, there are no exclusive conditions that all the business representatives should know how to speak English. This can be solved by using a bilingual approach for all shared resources, better planning and facilitating meetings where there are participants that use different languages for communication.

Different time zones – time management
A DD setup can also work with different time zones. Time planning and management is important, so the team can use the time overlaps in the best possible way as the project progresses. This should be considered at the project's earliest phases.

On-site support – proxy
Projects usually provide on-site production support for customers. This could have limitations if the team is distributed in locations far away from the customer or in other countries, and also from a legal perspective. This is solved by introducing proxy roles within the team that is closest to the customer. This role allows the whole team to be involved and contribute to the production support on their project.

Cultural differences and initial mindset/one-team feel – co-located kick-off, regular visits, distributed sessions and coffee breaks
If the team is distributed between different countries, it is important that they are aware of and understand cultural differences. Also, the initial mindset of one team can be challenging to achieve. It is a key

to a very successful distributed team. This mindset can be achieved by a co-located project kick-off, having regular bidirectional visits, and organising distributed sessions for knowledge sharing, pair programming, planning and decision-making. In this way, team members can learn about each other, get to know the cultural differences and, by understanding them, communicate and perform better as one team. Other powerful tools are informal distributed sessions such as distributed coffee breaks. Yes, it's as simple as that: grab a coffee or your beverage of choice, join the group meeting, turn on the camera and have fun – talk with your team members about anything you want :)

5. What can you distribute?

Almost any role that doesn't require close proximity to the customer can be distributed. We have experience of distributing roles such as PO, Scrum master, meeting facilitator, software engineer, QA, SD, UX, DevOps, PM, Agile coaches and many more.

6. What setup do you need for it?

Internet of course, and people :) Infrastructure is an important aspect that needs to be sorted out by having dedicated rooms/places for video conferences, good headsets, and other video and audio equipment. If properly set up, each of your teammates is only a click of a button away. Team spirit and motivation should develop as the project progresses, but there is nothing better than getting the whole team together at one location and defining a good onboarding plan for new members. It is also important to set up online tools and do all the necessary preparation (accounts, spaces, licenses etc.) at the earliest possible stage.

After all is said and done, with good preparation following the above guidelines, distributed development can be a very powerful asset to the company and a very fun and efficient way to run projects.

By Marko Ivanović

Links:

Your team needs a tech lead, not a lead techie

In what follows, let us assume that a tech lead is an experienced software engineer who is simultaneously supposed to lead the development team and be responsible for the entire technical solution.

The lead shock

There are several points that tend to come up when you ask tech leads about their career experiences, but the most common one seems to be that they were overwhelmed the first time they worked in that role. Why is that?

In my humble opinion, there are two key factors.

Firstly, stepping into the tech lead role brings with it an explosion of responsibilities that a software engineer has never experienced beforehand. In addition, many of these new responsibilities are non-technical and thus often very hard to grasp for someone with an engineering background. Becoming a tech lead not only includes obvious shifts such as switching from *moderate-scale thinking* to *large-scale thinking* or from implementation to concept work; it also includes switching from 90% hard facts to aspects like collaboration, communication, long-term risk management, expectation management, relationship management, etc..

Secondly, the skills required for these new responsibilities are particularly hard to obtain. That is because skills like technical foresight or the ability to detect misunderstandings before they cause damage come from experience. And in the same way that "experience is a hard teacher because she gives the test first, the lesson afterwards" (Vernon Law), it is also true that experience is hard to teach because no student in the world can comprehend the abstract lesson without having felt the concrete situations from which it arose. As a consequence, many organisations fail in preparing engineers for a tech lead role.

The recipe for dealing with the second factor is quite simple. Experience can be gained by assigning a future tech lead additional responsibility in small increments, ideally supported by coaching, mentoring and networking possibilities.

Let us now explore the first factor:

Responsibilities

Many authors claim that a good tech lead should spend at least around 30% of their time writing code. That may be helpful in many setups, but it certainly isn't in the ones I'm talking about. In the projects in which I am involved, writing code is a thing that the team already knows how to do well. As a tech lead, I wouldn't help them by doing more of the same. On the other hand, each hour I spend writing code is an hour I cannot invest in issues beyond coding — issues that need to be resolved in order for the team to make the most of their working hours. A tech lead should be a *multiplier* for the team, and *adding* code to the repository does not help in this regard. After all, it makes much more of a difference to help ten developers be 10% more effective than to contribute a mere 0.3 full-time equivalents of coding power.

So how can the tech lead make a team more productive? This is important stuff, so allow me to elaborate.

Developer productivity

For the sake of brevity, let us simplify things considerably by saying this:

The project manager and the tech lead define the development process, documentation guidelines and other general constraints.

The tech lead takes in the product vision, plus high-level requirements and cross-cutting concerns, and outputs a definition of the big picture, namely how the system is broken down into components, what responsibilities these components have, and how these components are supposed to collaborate.

Each developer is responsible for a couple of components. She or he takes in requirements that affect these components, plus architectural decisions, and outputs implementations of the respective features that adhere to the architecture as well as all the other constraints.

This may be a crude simplification (especially for mature teams), but it suffices to understand a crucial point: the productivity of developers depends to a considerable extent on the quality of their input, that is, on the quality of requirements, architectural decisions, process definitions, documentation guidelines, and so on. Therefore, a tech lead can make a huge amount of difference by making sure that the quality of this input is high.

Garbage in, gold out?

The last sentence can be put as a rule, too: the tech lead needs to act when the team is expected to produce high-quality output from low-quality input (in the above sense).

Here are three consequences of this rule:

The tech lead needs to team up with the project manager in order to help him or her define processes and guidelines that have a good balance between formal needs and everyday applicability.

The tech lead needs to team up with the business analysts and requirements engineers in order to help them produce output that will enable developers to process it efficiently.

The tech lead needs to define an architecture that allows developers to reason about the system despite its overall complexity (a complexity that, in its raw form, exceeds the capabilities of any single human being's mind, cf. Dijkstra's 1972 Turing Award lecture).

Drill-down

As the above rules and consequences are quite abstract, let me make things clearer by listing some specific lessons in this regard:

There is a human tendency when it comes to specification and documentation, namely a tendency to describe trivial and blatantly obvious things in minuscule detail while hardly even mentioning the complex stuff (because it hurts in the head). A good tech lead will watch out for this anti-pattern and react accordingly.

Software documentation in particular is often very unpopular because developers are (a) forced to work with inappropriate templates, (b) asked to document low-level details that are obvious from the code, and (c) not guided to document the overarching design decisions and calling conventions that are NOT obvious from the code (and that are thus really worth documenting). A good tech lead will make sure that the templates such as the templates for the software design documents for the individual components, encourage documenting the relevant, non-obvious information.

It is often the case that processes and templates are defined by people who do not have to work with them. That is a recipe for disaster since efficient applicability is not checked; instead, the overhead for useless work induced by the definitions can be arbitrarily large. A good tech lead will therefore intervene with all their might when they spot bad processes or templates.

Business analysts and requirements engineers may be good at judging the benefit of a feature, but they cannot be expected to be good at judging the implementation or maintenance costs. A good tech lead will help them understand the cost/benefit ratio by explaining the technical complexity in a comprehensible fashion.

In the same way, a good tech lead will act as an interpreter between the customer and the development team.

As for the architecture of the software system, a tech lead should keep the following rule of thumb in mind: developers implement features, and implementation is always a bottom-up process. Architecture, on the other hand, is a top-down issue: starting from the product vision, it defines a technical breakdown of the system into components and conventions that result in a uniform, intellectually manageable whole. Hence, there is a sweet spot: the tech lead needs to define the architecture down to a level that the developers can work against, but no further. If the architecture definition is too shallow (underspecified and abstract), then the developers will struggle because they don't know what to do. If the architecture definition goes too far (overspecified and too concrete in nature), the developers will suffocate in constraints that forbid them to solve their problems in their own way.

Similarly, a good tech lead will be aware of the following fact: a developer's work is about depth, not breadth. Many implementation tasks require the developer to dive into the code or even the silicon for hours on end; each distraction forces the developer to sort their mind and dive into the issue again from scratch. The tech lead's job, on the other hand, is about breadth, not depth. Therefore, the tech lead needs to provide guidance on the overarching aspects (module collaboration, lifecycle aspects, inter-component versioning, system integration, multi-threading issues, etc.). The boundary — what can the developer provide, what does the tech lead need to supplement — depends on the seniority of the developer, and the tech lead needs to be aware of that.

Furthermore, a good tech lead will have a sixth sense for technical risks and unexpected effort. For example, the software upgrade mechanisms and the configuration management for distributed systems are usually underestimated, and a tech lead will keep that in mind when estimating a project's cost.

Final remarks

Leadership is a very complex subject, and hence, it is a bold undertaking to try and describe how it works in a single article. Nonetheless, I hope that my perceptions contain something that you can put to use, and that I have managed to resolve the mystery of the tech lead question at least a tiny bit.

By Daniel Mölle

References:

PART III:
MACHINES, CODE

Application first – a bottom-up architecture approach

When advising a large company, we often encounter a scenario in which a range of small to large applications, written by employees for various purposes, has been developed. Commonly these tools are spreadsheets stored in Microsoft Excel with a portion of business logic or databases created in Microsoft Access. Those applications accumulate a significant amount of knowledge and information that is essential for the business. The apparent ease with which such applications facilitate the work of the business units ("Let me just do this in Excel") ensures a rapid spread of these applications as "undercover projects" and thus presents the company with a broad set of challenges.

The challenge

On the part of the business units, business-critical data and information are handled in an uncontrolled manner and this goes unnoticed by the company's IT department. This entails risks, not only in terms of data protection and the availability of the applications, but also in terms of non-compliance. Both may have serious consequences for the company. In addition, the business unit staff spend a considerable amount of their time maintaining these applications instead of concentrating on their actual duties.

On the part of the company's IT departments, however, there is a need to curb this "uncontrolled growth". To ensure that this can be done efficiently, in the eyes of both the employees and the company, it is important to ensure a standardized application landscape. This allows for new applications to be developed quickly and for existing applications to be maintained cheaply.

A bottom-up architecture approach

A popular choice is to first develop a common framework. By decree, this is then used as a foundation for all applications created henceforth. However, creating a framework is usually expensive and time-consuming without creating direct added value for the company's business. In addition, there is a high risk that the framework does not meet the actual requirements when the applications are eventually developed.

When facing such challenges, we usually guide our customers in a different direction: Instead of creating a framework upfront, we simply create the first of the new applications. This application not only delivers added value to the business right from the first release, but acts as a nucleus for upcoming projects too! We call this first application the "incubator application".

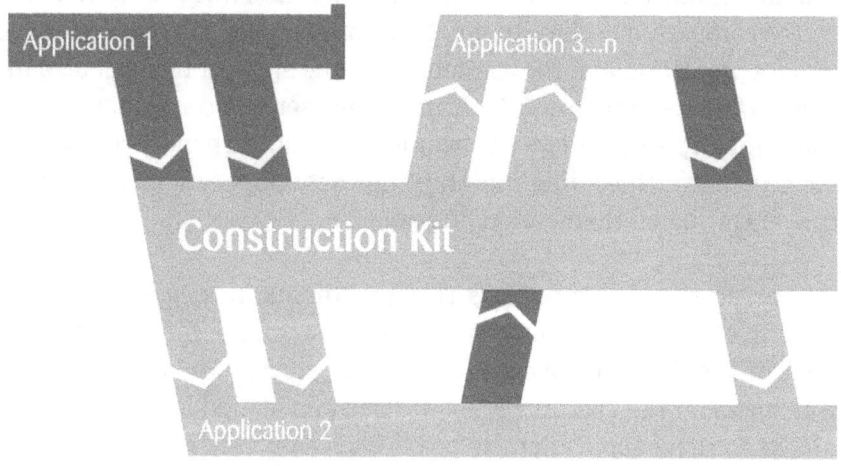

The development of the second application starts with generalizing components from the incubator into an "architectural construction kit". This includes the application skeleton with a basic UI layout, navigation structure and authentication and authorization capabilities.

As the family of applications built using our construction kit grows, so does the construction kit. While more and more components are shared between applications, it is important to provide the re-usable artefacts in an appropriate and easily accessible form, together with comprehensive documentation. This approach enables effective and efficient development, since the developers do not have to complete recurring standard tasks repeatedly with each project. On the other hand, it increases the quality of the software in the long run, since defects that have been fixed in the construction kit remove the issue from all affected applications.

Money for nothin' and change for free?

Although this approach is certainly cost-effective, it does not come for free. Both the construction kit and the developers' mindset need active maintenance. Otherwise, the once flexible construction kit will become stale and turn into yet another framework rather quickly.

Usually, the construction kit maintenance will be carried out as part of the ongoing application development. But as the number of applications and developers grows, it is essential to have an experienced developer acting as what we like to call a "free electron". This developer should have both the time and the budget to enable communication and know-how transfer between project teams. This ensures that synergies between the projects are reliably recognized and utilized, and that differences in the know-how of developers can be balanced by suitable measures. A question we usually get in this context is: who is going to pay for the free electron? In our experience it is important to fund this role in a way that does not strain the projects' budgets. This mitigates the risk of projects missing out on using the construction kit due to financial concerns. It also ensures that maintenance of the construction kit is not being rationalized away due to being seen as just a cost factor with no added value.

Conclusion

Using a bottom-up architecture approach has many advantages over a top-down one. Apart from delivering additional value to the business right from the beginning, it comes with a high return on investment: projects will get started faster with fewer problems. Due to a common set of components and common standards, developers will gain speed more quickly. Re-use becomes a reality and not a nightmare. Try it for yourself: If you start to think about developing a framework, start with a real application instead – your framework will emerge over time when needed!

By Markus Rehrs

Architectural programming

Architectural Programming (APRG) is a programming discipline for architectural elements and structures such as services, data sinks and sources or communication channels. It abstracts from infrastructure elements and hence is distinct from infrastructure as code. The approach requires APIs for architectural elements in order to create and evolve an architecture model using code. Azure, Google and AWS, for instance, all provide such APIs with services ranging from IaaS, CaaS to PaaS.

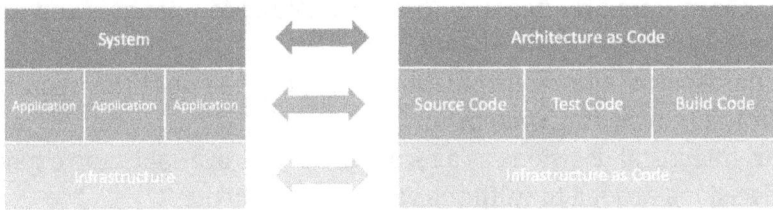

With the first implementation of the APRG approach, we extended Structurizr [Structurizr], an executable Architectural Description Language (ADL), with an explicit and coded relationship to the cloud infrastructure that is actually needed to implement and execute the system. This bridges a gap that nowadays still exists in most software development projects, a gap between models and code.

The coded model is the very basis for an envisioned overall development workflow that allows the validation of architectural decisions by executable quality attribute scenarios similar to the validation of acceptance criteria by automated functional tests. By expressing the model as code, compliance with quality attributes such as "all storage services are only available from virtual networks" can be tested.

Architecture models in the product lifecycle

Architecture modelling takes user and business requirements into account in order to provide a guiding structure and enable decisions regarding the implementation and operation of a system, which is ultimately delivered as a product to the customer.

Such a model comprises at least the following:

- system architecture for tiers, infrastructure and connectors between tiers
- application architecture(s) as a grey-box view of the different parts of the system architecture. This view details components and connectors between components
- quality attribute requirements specifying the qualities required from the resulting system
- architectural decisions as a record of options available and explicit decisions needed to fulfil the given requirements

In the following we consider mainly the system and application architectures. An extension of the approach to requirements and decisions requires further research beyond that which we have implemented so far (see resources: [Structurizr.InfrastructureAsCode]).

Architectural erosion and the model-code gap

Architectural erosion is the divergence of the architecture model from the source code and infrastructure that actually implements the model. It results in a model-code gap and usually happens gradually during the iterative and incremental development and maintenance of a system. The code and infrastructure evolves, but the architecture model is not kept in sync. After a while, we end up with a model that describes the system as it was meant to be in the beginning but not as it is actually implemented and delivered.

Such a system is often called "historically grown". It is difficult to maintain and to evolve, since decisions are often implicit, based on local knowledge, and no longer comprehensible once some time has passed. Fulfilment of the quality attribute requirements becomes harder and harder to achieve.

Towards Architecture as Code

Our approach is one step towards closing the model-code gap. It combines Architecture as Code with APIs for Infrastructure within the same code base. As an example, consider the following architecture of an Internet of Things solution built to monitor a factory producing stuffed animals:

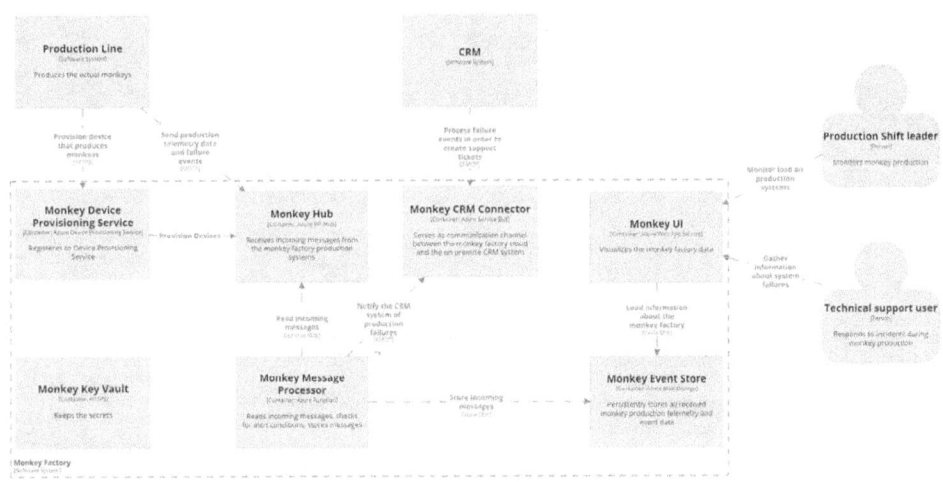

The frontend tier initially integrates directly with the event store, which is an Azure blob storage solution. After some development iterations it is decided to separate the persistence logic and turn it into an event manager component acting as a facade to the event store. An update of the diagram, taking into account this new component, is necessary and if not done leads to architectural erosion.

Instead of modelling the architecture in a diagram, we may use code like the following:

```csharp
using Structurizr.InfrastructureAsCode.InfrastructureRendering;

namespace Structurizr.InfrastructureAsCode.Azure.Sample.Model
{
    public class MonkeyFactory : SoftwareSystemWithInfrastructure
    {
        public MonkeyHub Hub { get; }
        public MonkeyMessageProcessor MessageProcessor { get; }
        public MonkeyUI UI { get; }
        public MonkeyCrmConnector CrmConnector { get; }
        public MonkeyEventStore EventStore { get; }
        public Person ProductionShiftLead { get; }
        public Person TechnicalSupportUser { get; }
        public SoftwareSystem MonkeyProductionLine { get; }
        public SoftwareSystem Crm { get; }

        public MonkeyFactory(Workspace workspace, IInfrastructureEnvironment environment)
        {
            System = workspace.Model.AddSoftwareSystem(
                Location.Internal,
                "Monkey Factory",
                "Azure cloud based backend for processing data created during production of monkeys");

            Hub = new MonkeyHub(this, environment);
            CrmConnector = new MonkeyCrmConnector(this, environment);
            EventStore = new MonkeyEventStore(this, environment);
            UI = new MonkeyUI(this, EventStore, environment);
            MessageProcessor = new MonkeyMessageProcessor(this, Hub, CrmConnector, EventStore, environment);

            TechnicalSupportUser = workspace.Model.AddPerson("Technical support user", "Responds to incidents during monkey pr
            TechnicalSupportUser.Uses(UI, "Gather information about system failures");

            ProductionShiftLead = workspace.Model.AddPerson("Production Shift leader", "Monitors monkey production");
            ProductionShiftLead.Uses(UI, "Monitor load on production systems");

            MonkeyProductionLine = workspace.Model.AddSoftwareSystem(Location.External, "Production Line", "Produces the actua
            MonkeyProductionLine.Uses(Hub, "Send production telemetry data and failure events", "MQTT");

            Crm = workspace.Model.AddSoftwareSystem(Location.External, "CRM", "");
            Crm.Uses(CrmConnector, "Process failure events in order to create support tickets", "AMQP");
        }
    }
}
```

Now, when this gets executed, the corresponding infrastructure and connectors are created and the system is ready to be used according to its description in the code.

Note that the implementation is based on Structurizr, an executable ADL which also allows architectural diagrams to be generated. Therefore, we may additionally generate up-to-date views of our current models, as needed, in the system's Architecture Guidebook, for instance.

In case we need to evolve or refactor the architecture, we work directly with the code. Only by this means are we able to modify the actual system and hence the model is always in sync with the code and infrastructure. Of course we could also change the system directly, but this is similar to changing code in a running system without updating the underlying source code in the repository, an anti-pattern which is hopefully nowadays no longer practised anywhere.

Finally, code can be executed and tested. If architectural requirements are implemented as tests, the architecture can be deployed to a dedicated environment and then the tests executed in a similar manner to functional acceptance testing. If we change the architecture in the next iteration the tests may fail, indicating the incompatibility of the architectural change with the existing quality attribute requirements.

Conclusion and outlook

The idea of Architecture as Code provides a required abstraction to Infrastructure as Code. With tools like Structurizr and its extension we move further towards executable ADLs which not only describe but actually *implement* the architecture of a system.

By Stephan Janisch, Christian Eder, Alexander Derenbach

Further reading and additional resources:

Architectural programming in the development workflow

Envisioned development workflow

How does the architectural programming (APRG) approach fit into the development workflow of a software product? We think the first important step is to have the model managed in the same repositories as the source code that implements the functionality of the system. This provides the very basis of a system development approach that allows architectural decisions to be validated and tested in a similar fashion to the way in which behaviour-driven development [BDD] works for functional acceptance.

Version control and executability

Most development teams document everything in a wiki (e.g. Confluence). As this wiki is often the central place for documentation, it feels natural to store the architectural documentation next to the business documentation. In many cases, this documentation also needs to be delivered to the customer, so having it in one place seems a meaningful approach. But from the perspective of evolving architecture, it makes much more sense to store the documentation with your source code. It could be stored on the same branches and tags as the source code without the need for any extra housekeeping. In addition, less distance between them will support keeping the gap small.

Architecture is often expressed in diagrams. On the one hand, they can give an overview of the system while, on the other hand, you can dig as deep as required into the details of a system without writing dozens of documents no one ever reads. Many current tools already use an XML or JSON format which could be checked into a VCS. But anyone who has ever merged conflicts in an XML document knows that this is not the best thing you can do. Additionally XML and JSON formats provide no (or at least less) semantics which could be executed. And often multiple diagrams show different views on the same model. Every change of the model requires a change in all views.

Using architecture as code is beneficial here in all cases. Merging the code is much more intuitive, as the language can be the same as the source code itself. Also changes in the model are type-safe (depending on the language you eventually use at execution time), so all views will be taken into account.

And, finally, code can be executed and tested. Architectural requirements which are expressed by code can be verified. The code can be processed in a CI/CD pipeline. After running the tests, the architecture can be deployed in a dedicated environment.

Decisions, scenarios and tests

A further step towards bringing the complete architecture model and the code together is to store your architectural decision records (ADR) with your code. ADRs are short text records in a defined format. This helps to keep decisions simple but informative. See [ADR] for further details.

ADRs are made to fulfil quality attribute requirements for a system. So ADRs are the architectural stories of the system and quality attribute requirements are the acceptance criteria of these stories.

We could write tests based on these acceptance criteria to prove which ADRs are fulfilled. As described in [EVOL], writing these tests can be a challenge, since there will be no one recipe to rule them all. One test could involve the execution of a performance test in a special environment set up by the architecture code to test performance requirements. Another could involve automated penetration tests for security or resilience tests following the principles of chaos engineering.

The important thing is for the test system to be set up using architecture code. Only then is it possible to prove that the system still satisfies all requirements. Since the effort for testing must, as usual, be balanced against the actual value it creates, it is advisable to follow a risk-driven approach and create ADRs and tests only for the parts of the system where the potential risk indeed requires it [RISK].

Conclusion and outlook

The current state already provides a good basis for storing architecture as code with your source code. With ADRs you can already store the important decisions under version control. Additionally the approaches regarding architecture as code get increasingly mature. Implementing the architecture gives you the full spectrum of software engineering techniques for your architecture, starting with version control, CI & CD pipelines and testing against given acceptance criteria. Parts of this idea already exist, others must be developed and some still sound like fiction.

By Stephan Janisch, Christian Eder, Alexander Derenbach

Further reading and additional resources:

CI and CD done right

Short introduction and history of CI

Continuous integration (CI) was adopted and driven by the extreme programming (XP) methodology in order to combat *integration hell*. XP first advocated writing unit tests which every developer can run locally before merging to the main line. In later iterations of XP, the concept of a build server was introduced and further improvements led to what we refer to as CI:

- Fast automated builds
- Run on every commit
- Including tests
- Run by some mechanism that can provide feedback to developers.

Short introduction to CD

Continuous delivery (CD) builds upon CI with the aim that the mainline branch can be released and deployed to production at any time. Continuous delivery is quite similar to continuous deployment but continuous deployment will deploy any merges to mainline to production (if tests pass), whereas with continuous delivery, releases and deployments are triggered by a human. For an example see below:

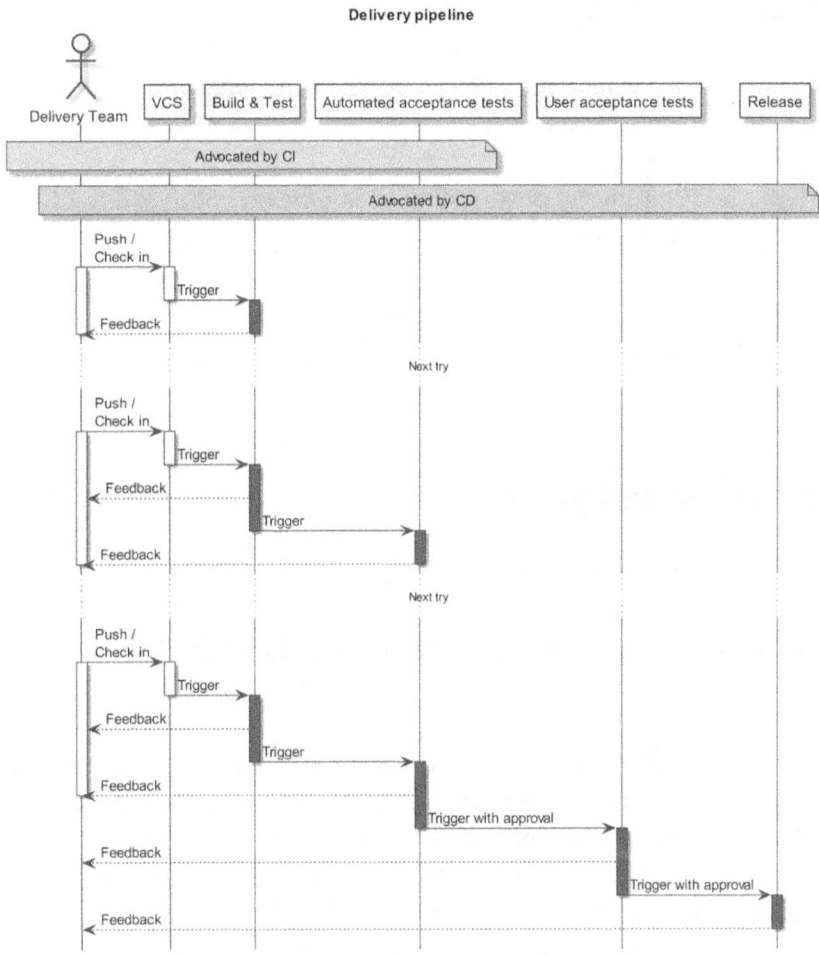

The pipeline follows the *fail fast* approach, delivering quick feedback and ensuring target systems remain in an acceptable state if tests fail.

How to do it right

There is no *one way* to do CI/CD right. The same way you can't have a *right* way to do Scrum or XP.

There are a number of obstacles to overcome on your road to CI and CD. Obstacles you must overcome can be categorised as:

1. People
2. Organisations
3. Tooling

Issues around people

Inexperienced developers

They tend to write code that is not designed for testability, resulting in

- fewer tests, which then results in bad quality and an increasing chance for a bug to sneak into production. This can torpedo CI/CD because people advocate "taking more time" to fix issues. It is important to remember that CI/CD can give you the same quality *faster*. Your best approach will be to coach or lead by example on how the issue could have been fixed without sacrificing speed. Your success depends on how much the team believes in CI/CD.
- Tests using a lot of mocks tend to be more fragile. Fragile tests directly hinder integration because tests have to be fixed. Here, your best bet is enabling developers to design better code so they can get rid of fragile tests altogether. It is imperative that you act early because once this issue escalates, CI/CD will be seen as a constant pain.

- Slow/flaky tests with a difficult setup on a higher unit than necessary. Your delivery pipeline slows down and must be fixed by demonstrating better software design to your team.

Large product backlog items

Large product backlog items (PBI) or big impact changes created by inexperienced requirements engineers are the enemies of quick integration if they must be integrated as a whole. Inexperienced developers may be tempted to defer integration until "their" PBI is done, leading to integration hell as by then a sizeable piece of the codebase will have been changed.

Scepticism

A lot of people are sceptical until they experience the benefits. The best way is to focus on a small set of pain points and to address them transparently. There may be people that object to the idea of having a faster integration cycle fundamentally; if their beliefs are motivated by personal reasons then you may need to restructure your team. Examples are:

- Clinging to manual tests out of fear of losing control
- Weak confidence (without explanation) if build is green
- Taking pride in a "build master", that is a human quality gate through which all code must pass

Organisational issues

- Increase in pressure: An organisation may redirect its resources away from CI/CD. Prepare stakeholder buy-in before starting CI/CD, be transparent and point to buy-in throughout implementation. There is no shame in failing transparently.
- Unsuitable process: There may be other processes conflicting with the goals you want to achieve. Get management buy-in beforehand and pick which processes to challenge; you will lose some battles to win the war.
- Unsuitable infrastructure: It may be very difficult to realise CI/CD with the current infrastructure. This is something you should check before starting and raise as a precondition. Explore different infrastructure providers by going to the cloud, or request new hardware.
- Code review ping-pong: Asynchronous code reviews take longer and require more task switches. Try out face-to-face reviews or pair programming.

Tooling issues

- Slow regression feedback: You may experience slow feedback due to hardware constraints. The first solution is to pay for better hardware (which is cheaper than humans). Otherwise see "Slow/flaky tests" above.
- High-noise feedback: The feedback from your pipeline can overwhelm people, and they will ignore feedback altogether. Make sure you only send feedback in cases where human intervention is required.
- Long integration queue: If you see a build server being overloaded with requests, get more servers/scale out in the cloud.

Solving obstacles around people is both the hardest and most important challenge you must overcome, while tooling is the easiest and least important. A successful CI/CD pipeline will make you and your team more productive and will increase quality.

By Florian Besser

Further reading and additional resources:

Clean code best practices

Being a developer these days involves sometimes developing features and fixing bugs at a fast pace, which can result in code that is difficult to maintain in the long-term. In other cases, we want to make the most elegant solutions possible to solve some problems. This can produce code that machines can understand but people have more difficulty understanding. Such code can become a major issue for a company to maintain and use in the long run.

Our objective should be to write maintainable, understandable, simple and readable code, and we need to make an extra effort for this to happen. It should be easy for our colleagues to change and understand the code we create, but this is not an easy task to achieve, especially in large projects. So we need to practice writing tidy code and fail, fix it and repeat again until we achieve clean code. We can do this by re-writing code as we go, rather than waiting for big refactorings. Another good approach is that we try to type out all our code, rather than copying and amending it, as we often do, as the former results in a better understanding of what we are doing, which helps us produce better and cleaner code.

This is the best way of creating clean code, and it also helps to apply some guidelines, principles and techniques, as described below.

- Naming – Naming things is one of the most important things in software development. Names are everywhere in software. We name methods, classes, files, etc. It is important to give meaningful names as they need to indicate the purpose of that part of code. Names should be expressive and clear enough to allow us to immediately know what the code does. This also helps with the implementation of a self-documenting approach.

- Clearness – As a main guideline, the problem that the developer needs to solve is of critical importance, but the software solution must be understandable for a developer who didn't write the original code.
- Focus – Good code should comply with the Single Responsibility Principle (SRP), so that the code we write has one specific purpose and is compact, as well as being responsible for a single part of the functionality.
- Simplicity – We should always try to apply the DRY (Don't Repeat Yourself) rule. What this means is reducing any repetition of code, so that a single change does not require numerous changes in the code. In addition, we should follow the KISS principle (Keep It Simple Stupid), which forces us not to complicate things unnecessarily.
- Readability – In order to make code more readable, we can follow the YAGNI principle (You Ain't Gonna Need It), which implies that we shouldn't write code upfront that is not currently needed.
- Commenting – As comments usually represent an anti-pattern, we should be careful with them. If we insert a lot of comments, this usually means that the code is not self-documenting. We should instead focus on creating more clear and readable code and not having to add numerous explanatory comments in the code.
- Formatting – We always first tend to focus on creating software that works, but in the long run we also need to have good readability. To achieve this, our code needs to have a coding style that is understandable by a team, because otherwise maintainability can be seriously affected, especially in large projects. Although we now have powerful tools that can take care of formatting, teams should still adopt formatting rules and follow them.

Along with these guidelines, a developer should use "code sense" when looking at code. This allows people to see different options, so that they can select the best variation of it, to make value-added and clean code. To do that, we must practice coding and explore different dimensions of programming languages.

It is also important to remind ourselves that we need to write nice and clean code, but also code that solves the problem. Of course, we need to deliver code, whether it is clean or not, rather than not delivering it at all. However, the creation of clean code should always be a priority, as this will bring more benefits to a project in the future. It should become standard practice to go back and clean the code before moving on to the next task.

By Milan Milanović

Codify your developer VMs!

Let's talk about developer VMs, and why you absolutely and definitely must automate them (no excuses)! :)

What is a "Developer VM"?

A developer VM is a virtual machine image that contains the complete development toolchain that is needed to work on a specific project (including compiler toolchains, IDEs, system wide settings, etc…). The aim is to ensure a consistent environment not only between team members of that project (which may have completely different operating systems on their laptops), but also with the build agents in the CI environment.

Why would I need a developer VM anyway?

Some people hate it, others love it. However you put it, it is a sure-fire way to ensure that the whole development team uses a consistent environment and avoids the typical "works on my machine" issues we all know about. Especially for us, executing and delivering a multitude of projects (and engineers joining and leaving these projects more or less frequently), developer VMs have become an indispensable tool. Not only for the consistency aspect, but also for reducing ramp-up times or being able to easily archive the development environment when the project enters the retirement phase.

On regulated projects this is even more relevant, since you need to validate and verify the development toolchain and need to ensure that everyone uses exactly that defined toolchain.

OK, but why should I automate it?

While we started out to craft these VMs manually (and with love) in the early days, we can nowadays easily automate the whole procedure with tools like Vagrant, Chef, Ansible, etc. Applying automation and Infrastructure-as-Code principles to developer VMs not only increases transparency and reproducibility, but also allows for an in-VM update mechanism to maintain consistency while the toolchain evolves during the lifetime of the project.

Finally, an automated developer VM should exhibit the following properties:

- updatability – an existing VM can be updated with the push of a button from within the VM
- testability – automated tests are run in the VM to ensure that it works as expected
- installability – the VM is distributed as a standard VirtualBox / VMware image
- adaptability – developers can easily adapt and improve the VM as the toolchain evolves
- reproducibility – changes to the development environment are transparent, traceable and reproducible

How can I automate it?

The good news is that there are quite a few so-called "Configuration Management" tools out there, which essentially allow you to automate the provisioning of software and configuration on top of an existing operating system (in our case a fresh installation of the operating system of choice, within a VM image).

The bad news is that you have to choose one. In addition to the configuration management tools that do the provisioning, there is (fortunately) also a multitude of tools for testing the configuration of such a (virtual) machine.

Depending on your preference, these are the tools you are likely to end up with:

Rubyists will like
- Chef – as the configuration management tool
- Foodcritic – for linting their Chef recipes
- InSpec / ServerSpec – for testing the Chef-provisioned systems

Pythonistas probably prefer
- Ansible – as the configuration management tool
- Ansible Lint – for linting their Ansible roles
- TestInfra – for testing the Ansible-provisioned systems

It is not all a bed of roses though!

Two things to highlight:

1. Developer VMs are clearly a compromise between consistency and performance: you save valuable time by avoiding the "works on my machine" issue and speeding up onboarding time for new team members. The price you pay is the runtime overhead for running the VM in a hypervisor rather than working with the tools natively on your operating system of choice.
2. The automation part is sometimes hard (especially on Windows) and is not just a case of software installation but also configuration of settings like networking, firewalls, access rights, etc. These are not things that everyone knows how to automate, and

you will have to grow and master these skills as part of your journey.

I'm still sold, sounds like pretty cool stuff! What is the fastest way to get there?

It may sound like a bit of effort is required to achieve this if you start from scratch, so that is why we have a template / skeleton project along with a tutorial for you. No excuses for doing it manually anymore! ;-)

This is what you need to get started:

- A ready-to-use template / skeleton project for a minimal, Linux based developer VM: https://github.com/Zuehlke/linux-developer-vm
- An example Java Developer VM based on the above (see Pull Requests for a step-by-step guide): https://github.com/tknerr/etka2017-developer-vm
- The accompanying conference talk for creating the example above: http://bit.ly/automated-developer-vms
- A ready-to-use developer VM to with a toolchain for automating developer VMs: https://github.com/tknerr/linus-kitchen

By Torben Knerr

Containerisation and why to use it

You are a good software engineer, but releases are a time of pain. Your project matches one or more of the following:

- Deployments are done manually or via scripts that execute on some remote host.
- Scaling the app is hard as new machines or VMs take weeks to arrive and set up.
- Your application dependencies differ in version or feature set between environments.
- Developers have difficulty reproducing production issues on their local machines.
- Application dependencies are installed manually, developers spend days setting up their own machine.

This article will explain how containerisation can help solve each of these issues.

Cumbersome deployments and dirty hosts

Running scripts on a host seems OK in the beginning, but over time your host will become "dirty". Some versions of your script were only run in QA, and production has leftover artefacts from previous deployments. These leftovers confuse maintainers and reduce confidence in your team's ability to deploy software to production systems.

With containerisation, you immediately get rid of leftovers your software may cause. By throwing away the old container after deploying a new version, you have already cleaned up after your application.

Sometimes you want to persist data across container crashes or restarts. In these cases you can use volumes to store the data and not pollute your host. Volumes allow you to use containerisation with databases quite nicely: The data is safe from crashes but up- and downgrading a DB is just a question of stopping the old container and booting a new one.

More hardware and the procurement process

There is a conflict between needing more resources and a company's willingness to spend money. Usually, there is a long-winded procurement process for ordering hardware and getting it installed in a rack. At the current time, when hardware is cheap and humans are expensive, paying some PaaS a few hundred bucks a month is cheaper in most cases.

Any established platform will offer you support for Docker, and if you have containerised your application you'll be able to solve the scaling problem a lot more easily. Usually, this involves uploading the image to the platform, creating a service and then defining scaling rules for said service. While the platform will not run on your local machine, this does not prevent you from testing locally, as you can install Docker and are guaranteed the same runtime as on the platform.

Protip: Most platforms offer some functionality for doing rolling updates, you can deregister your containers from the load-balancer before killing them, thus achieving deployments that are invisible to the customer. Now every day is a production deployment day!

Different environments, different dependencies and the highway to hell

Differences between environments should be as small as possible, but they tend to increase over time. Someone upgrades only the QA database for testing; someone applies emergency security fixes only to production because other environments are not critical. You find an increasing number of builds pass QA but fail in staging or production!

Containerisation can remove this problem as whatever is containerised is versioned and there are orchestration tools that ensure all dependencies are the correct version. Orchestration means that the app and its dependencies are deployed together as one coherent bundle. This way you can remove uncertainties for everything you can containerise.

Note that many things a novice deems uncontainerisable are actually containerisable: From any database, including Oracle, to messaging systems to webservers, you'll find premade images or you can roll your own. Anything that runs on an unmodified Linux kernel can be containerised!

Production bugs and developer madness

We task developers with reproducing production bugs locally and fixing them. This approach can suffer greatly if the developer has to emulate production instead of running an exact copy. In the worst case, this leads to unreproducible bugs and some "blind" bug-fixing in the hopes that changing some related thing will solve the issue.

Luckily, with containerisation, we are able to run an exact copy of the version that is in production. If bugs are not reproducible, they depend on application dependencies (see above) or on the production configuration (but that should be easy to debug!).

Setting up ::1 and the onset of despair

As people join there'll be a recurring pain setting up new machines. Done manually it has two downsides: Any newcomer must experience the *setup* of all the dependencies used and they end up with different versions than their teammates. This introduces "works on my machine" between developers, costs a lot of time and demotivates newcomers.

But if you already containerise your dependencies and have an orchestration solution, then you run the orchestration locally. You'll get the correct dependencies, the correct version, and the correct configuration.

Note that some people argue that setting up everything yourself is a good exercise but I think the dangers outweigh the knowledge gain as the knowledge can more easily be gained by giving newcomers an overview and some first tasks with purpose.

By Florian Besser

Do something about that slow SQL query

Please, don't blindly buy more memory, faster disks or CPUs. To start solving database performance problems, first of all, you need understanding. So let's start removing the veil of mystery from the topic of SQL performance. Let's understand what can we can do to make our database run faster.

A typical performance problem would be one large query executing for a long time, or an enormous number of micro-queries bringing the database server to his knees.

First of all, you need to understand the cause of the problem. Do not optimise blindly. Find a way to monitor all the processes on a troublesome server. Be sure that there is no other process that is suffocating the server, i.e. that the slowness of the database is not caused by some external factor.

If the database is what is exhausting most of the server resources, you need to find out why this is happening. The database essentially responds to a set of queries sent by clients, using the data stored in the tables. You need to have an insight into these queries to know which one of them precisely is problematic. The database, or the ORM framework, usually has the ability to record all the queries in the log file, as well as to measure their performance. It is vital that you are able to search the log files, find the number of queries at a certain time interval, and find the slowest and/or frequently repeated queries. Sometimes for these things, it is sufficient to manually review the log files, and sometimes you will use tools (log analysers) to dig and visualise this information.

If the database is slow because of an unexpectedly large number of small but quick queries, then it is usually necessary to:

1. Change the client so as not to send an unreasonably large number of queries, but to join more queries into one. A typical problem with the ORM framework is unnecessary lazy loading. In case of reading multiple records from the database, this becomes an "N+1 problem": your query will get only basic data for all N rows, and the ORM framework will execute one additional query for every single row. Carefully selected eager loading (either as part of a query using the "fetch join" construct, or declerative) can significantly reduce the number of queries and improve performance.

2. If smaller queries are really necessary, then analyse the queries. It may happen that a lot of queries like ID/username lookup are being repeated over and over again. If that is the case, consider caching the results of frequently executed queries (either by using the ORM cache, a 3rd party cache, or rolling your own). It may speed up total query execution time dramatically.

If the database has one very slow query, or several of them, then they need to be analysed individually and possibly accelerated by the following means:

1. Create indices over appropriate columns (primary keys, foreign keys, columns in JOINs, columns for search / selection)
2. Create partial indices (e.g. over active records, or over records of a particular type)
3. Avoid selecting unnecessarily large amounts of data (e.g. selecting all the rows from a big table when the user might view only the first couple of pages, or selecting all columns when only some of them are presented to the user)
4. Optimise correlated subqueries

5. Check the plan for query execution and database statistics, and update statistics if necessary

Finally, if the number of queries is within expectations, and the execution of each query is slower than expected, there may be a systemic problem and the solution is to check the configuration. E.g.:

1. Not using the connection pool can slow down any query, because establishing a TCP connection takes time
2. Incorrectly configured memory allocations can contribute to unwanted swapping of virtual memory between RAM and disk, which is disastrous for performance
3. The wrong file system and RAID configuration (RAID 10 or several RAID 1 groups perform the best; avoid journalling on log partition file system; avoid RAID 5 as it does not perform very well).
4. Wrongly selected storage. For example: NAS and SAN, despite similar acronyms, are very different storage solutions with completely different performance.
5. Not enough RAM. If the entire database (together with indices) exceeds the available memory, performance may be poor. In the case of large databases that cannot be completely stored in memory, it may be unavoidable, but in the case of smaller databases, the biggest performance jump is obtained if the whole database is in the memory.
6. Bad hardware. Sometimes you just need to get better hardware.

Here, we have introduced some practical solutions for how to tackle database performance issues, and this can be used as a starting point for you personal optimisation quest.

By Ognjen Blagojević

Frontend is not your enemy

For many years now, we have been neglecting frontend web technologies.
- *"It is too volatile!"*, I've heard someone say at the back
- *"But I only learned Angular 1.0 yesterday, and today it's already v3.3.1"*, one Java developer yelled while hiding the tears...

We have been witnessing an explosion in frontend for a decade or so, due to so many innovative requirements to create the most user-friendly web app. As you will see, many good things have come out of this pile...

ECMAScript and OOP

ECMAScript is a standard that drives all new native JavaScript features. ES6 in particular brought many improvements that developers had craved for a long time.

A big step forward are classes. There is no longer a need for that cold sweat before we use Prototype or function constructors, hoping chaos won't unfold in the process. Classes are especially useful for developers who come from an object-oriented language like Java or C#.

There are also other features that you will fall in love with. Like *let, const, lambdas, destructuring, async/await, default params...*

I recommend taking a look into the ES6 documentation and getting familiar with the features.

Compile time checking

Ok, I guess that some of you might feel more comfortable working on *backend* with strongly typed languages, and you are used to having your IDE insult you with *Cannot assign boolean to string!* So you perhaps find frontend rather repulsive.

Well, guess what, you have all that now with TypeScript! You can even customise the guidelines and allow the team to agree on these, so that engineers will be notified if they, for example, define a method without a return type.

It has all the OO principles that Java has (and much more), such as interfaces, abstract classes, enums, etc...

Build tools

Yes, there's a proper build in frontend. And this has been the case for a while now. And do you know what, one can even import .css into .js files now! I know, what kind of monster would do such a thing. Well...

Node basically opened the door to a whole new universe of wizardry we can do in frontend development. Node is a popular JavaScript runtime built with V8 engine (the same as Chrome). With Node, you can easily build scripts or backend web APIs that execute on any OS environment. Many big players use it, such as Uber, Netflix, PayPal, etc... It started with simple task-runner packages (Grunt and Gulp) that can help you organize your code, clean-up, uglify/minify it, and much more. Now we have Webpack, the most popular build and bundling tool on frontend. Using various loaders, you can easily import modules and do all kinds of transformations on your code, from development to deployment phase.

When you have to choose a build tool, choose the one which is battle tested, has a great community support, is feature-rich and performant.

Frameworks and libraries

What kind of IT article would this be if the word *framework* didn't find its way on to the page somewhere. However, I will try to be unbiased and talk about them in general.

Should you use them? Yes! They help a lot with development and in terms of keeping your architecture clean. They also help with introducing new team members, as they become familiar with them. Most of the mature ones provide similar solutions and follow similar patterns:

- Data binding
- Routing
- High performance templates
- State management
- Virtual DOM
- Testing utilities

"*But which one should be used?*" Well, it is indeed painful to choose the best framework in the ocean of great ones. Currently, I would recommend evaluating React, Angular and Vue.

Testing

As frontend apps are becoming more and more complex and heavy with logic, it is crucial to treat them like grown-ups. Thus, testing JavaScript code is not a luxury any more, but rather our duty. Again, *"WHICH ONE?"*, you might scream, and again, *"It depends"*, I shall respond.

As you might have guessed, each framework has its own specific testing tool. These are some popular ones:

Jasmine
- Simple to setup
- Mocking support
- Widely used with Angular

Mocha
- Most used nowadays
- Very flexible
- Needs some additional setup time

Jest
- UI testing with snapshots
- Becoming increasingly popular
- Best choice when using React

Karma
- Flexible
- Run tests in a page-like environment
- Setup can be somewhat cumbersome

Other tools you might want to look into include Protractor, CucumberJS, Ava, Chai, Enzyme.

Final words

Either for your personal development tasks or for project needs, you should invest effort in mastering frontend technologies.

> *"Any application that can be written in JavaScript will eventually be written in JavaScript." – Atwood's Law*

By Janko Sokolović

How to deal with flaky system tests

The higher we climb up the test automation pyramid, the less oxygen there is in the air. Some tests are fine with that. But others can't handle it and randomly fail every now and then. These poor tests need help. In this article I will focus on strategies to keep these tests as stable as possible.

The problem with flakiness

A test that changes from red to green or green to red without any code changes is considered flaky. Such tests are a burden for several reasons. They might block you from merging to the mainline or to release a new version of your product. They might be a symptom of an unstable test or a problem in the product. In the end, dealing with all these issues eats up valuable time. That's why we want to avoid flaky tests as much as possible.

Increase your chances of success

Avoid broken windows

As soon as one test fails, insist on fixing it. Otherwise people might start ignoring failures and you'll end up in a broken window situation. There's a root cause to every flaky test. Either it's the test itself that needs improvement or it's a bug in the software.

Control as much as possible

The more control you have over the system under test (SUT) and the test data, the higher your chances of getting the tests stable. There are many ways to increase control. Here are some important ones:

- Preferably use system tests that cover only the system you are responsible for (at runtime) rather than system integration tests that also include the systems of other teams. This requires simulators to mock the other systems out.
- Define the exact versions of the libraries you are using in the code, so that you don't get surprised by unintended updates that introduce issues.
- Control the test data setup and automate it as much as possible.
- Make sure tests are independent of each other (i.e. one test must not rely on the outcome of another test).
- Increase testability by modifying the system code (e.g. by adding automation IDs to the UI).
- If your tests also cover other teams' systems, know when they do their deployments or have downtimes.
- Keep continuous health checks separate from your test suite; they should not have an impact on your build pipeline.

Timing is (almost) everything

Most issues involving flaky tests are related to timing and getting it right is often hard.

- It's crucial to know when the testing framework implicitly waits for an action to complete and when you have to add explicit wait code.
- Instead of fixed wait timeouts (e.g. 5 seconds), implement polling that checks in a loop for a certain condition to become true. This is much more stable.
- If your system under test has asynchronous behaviour, make sure you understand it well and write the tests accordingly.
- Configure sensible timeouts.

Analyse failures systematically

Reproduce failures locally

If you have to rely on a CI server and test environment to reproduce a failure, your feedback loop after every change will be very long. This makes fixing the issue a cumbersome and inefficient procedure. Therefore make sure you can start the test locally with the debugger attached. The closer your CI environment matches your development system, the higher the chances you can reproduce failures locally. If the test does not fail as often as on the CI, consider running the test in a loop to provoke a local failure faster.

Unless your development machine has exactly the same setup as the CI server, some failures will not be reproducible locally. You will be forced to analyse them on the CI environment. One of the build nodes could temporarily be used to analyse the issue while the others can still be used by the rest of the team.

Collect data from many runs

If you have to deal with flaky tests that fail very rarely, running the tests a few times a day might not give you enough data for analysis. Therefore, schedule the tests to run as often as you can. In a recent project, we ran the test suite once every hour throughout the night. This gave us a lot of data to analyse stability the next morning.

Aggregate results from feature branches

You might actually have a lot of data from running the tests on each feature branch, but it is hard to see patterns if you don't have the test runs in an aggregated form. Collect all the test failures from all branches in a single place to analyse them. You might start off with an Excel sheet, but will probably soon decide to automate the data collection.

Visualise

Visualise failures in a matrix where one dimension is the test run and the other is the test scenario. This might reveal patterns that can give you a hint where the problem lies. For example, if a certain group of tests always fail together, you might want to check what they have in common. And if a single test only fails every 10th time or so, plotting it on a graph will make it apparent that there's a problem with this test.

Analyse logs and metrics

Logs and metrics from the CI server that runs the test and from the SUT might give you valuable hints as to the root cause. If your servers run in VMs, you might also want to collect metrics and log files from the VM host. Collect everything in a central place so you can analyse the timeline of events and look for recurring patterns.

If all else fails

Rerun failed tests automatically

If you can't tame the beast or if you need a quick fix to buy you some time for a proper solution, you might want to rerun failed tests once or twice automatically (in Java you could use a JUnit rule for this). If the rerun is successful, the build is successful. This will make your build runs a bit slower, but might spare you a rerun of the entire build pipeline. However, this approach might have the undesirable side effect that people care less about writing stable tests as flaky ones don't hurt so much any more.

Delete it

If, despite all your efforts, you find it impossible to get some tests stable enough, consider deleting them. Think about how you can replace them with other risk mitigation strategies. Maybe a unit or integration test can cover most of the risk. In some cases you might need a scripted manual test scenario. Or you may find out that it's already covered by the exploratory testing that is done before each release.

By Adrian Herzog

Making your tests run fast

Automated testing of software is generally a good thing and is accepted as a standard practice in virtually all projects these days. However, more often than not, people get annoyed by the tests over time. They keep complaining about having to wait for those darn tests to finish on the CI, so they can finally merge their PR.

Clearly, no one sets out with the intention of slowing everyone down. But then again, people rarely make it a priority to keep things fast. Which is understandable and sounds "pragmatic" when there are only a handful of tests. Fast forward six months and people are sitting idle, staring at the progress indicator of the build pipeline, hoping that they did not break another test and will not have to waste another 30 minutes of their life.

So, what should you do instead?

There are a few strategies and most of them don't incur any additional cost – if done right from the start!

Don't write tests in the first place

Depending on the technology and programming language, there are ways to prevent certain errors by taking advantage of the compiler/type system or static code analysis. This gets rid of tests which need to verify behaviour at runtime that can be checked at compile/build time. A simple example would be using an unsigned integer type instead of a signed integer one if negative values are not allowed, but it can and should be applied to more complex scenarios as well.

Don't repeat yourself (too much)

If a particular functionality has been tested once, there is no need to have it tested again and again. There is also no value in testing a bunch of random cases instead of only one. Only add additional cases if the additional case proves some additional property of the system under test. Of course, there will be some overlap between tests, especially between the different levels, but it pays to ask yourself for every test if it really adds valuable insight. Not only will this avoid additional execution time – which can be significant for integration and system level tests – but it will also make the test suite more maintainable, as you don't have to change many tests if a requirement changes.

Get time under your control

The first time you need to implement time-dependent functionality, ensure the progression of time can be controlled from the outside. Provide an abstraction over any time-related primitives you might be using (getting the current time, waiting for time to pass, ...), which can be controlled from your tests to make time progress exactly and instantly when you need it to. That allows tests which check for, say, a 20-second timeout, to execute instantly and make testing of longer time spans (hours, days or years!) feasible.

Separate, separate, separate

Be clear about what properties are tested on which level (e.g. unit, integration, system) and keep those levels separated. Having clearly separated tests, "horizontally", allows you to define different guidelines, constraints and approaches to each level and makes it easy to run them independently. In addition to this "horizontal" separation, tests and their corresponding test subjects should also be separated "vertically" by applying the various techniques of modularization.

This enables you to run tests only if the corresponding code has been modified.

What if people are still complaining?

Even after you have done everything you can to make running your test suites fast, it likely won't be enough. Be it the sheer amount of test cases or the requirement for a lot of integration and system level testing, at some point things are going to be "slow". But there are some additional strategies for dealing with that:

Parallelize

Run test cases and suites in parallel, as much as possible. Do not shy away from distributing the execution across multiple (virtual) machines. Additional design and/or refactoring may be required to make that possible. However, if you have already applied all of the points above, you should usually be able to parallelize quite a bit without changing anything.

Prioritize (and fail early)

Categorize/annotate all tests based on how essential they are and how quick they are. The default should be for a test to be "most essential" and "fastest" and mark every test that deviates from that, e.g., exhausting options, testing more (theoretical) corner cases or more complex and slow workflows. Order all from "most essential" to "least essential", then from "fastest" to "slowest", execute them in that order and stop on the first failed test. While this does not help when every test passes, it helps to stop wasting time and resources when something fails.

Prune

Identify and remove tests which are not covering actual requirements anymore or never covered any particular requirements and are just testing implementation details. Just remove those tests. While this is not always easy and requires a good understanding of the system and its requirements, it is worth considering from time to time.

I don't have time to do all of that. What should I do?

If nothing else, keep an eye on test execution time. Measure it, visualize it and make people aware of it. Listen to people around you, if they start complaining about it. Just being aware that this might become a real problem in the future is often enough to plan for it right from the start.

By Simon Lehmann

Optimization and realization

There is no developer that hasn't heard the famous Donald Knuth quote: "Premature optimization is the root of all evil." Often this sentence is understood in the context of performance, i.e. the speed of the software execution. However, it refers more to something else.

It's about *the value* that some code brings into the product.

During the development, the programmer's goal is to write the code that effectively meets the required functionality. However, often in the desire to write a high-quality code, a programmer goes too far and introduces complexity that is greater than its real worth (over-engineering). In other words, the value of the code decreases because of the unnecessary increases in complexity. Similarly, sometimes the development focuses on features that are not critical and do not give the product essential value. Instead, development tackles less relevant features.

In this context, premature optimization is all the work that was not spent on the production of real value. The alternative to optimization is realization: work that actually brings value. Now the Knuth sentence is more meaningful.

Striking the balance between optimization and realization is not reserved for planning and top-level architecture. It make sense to view work on everyday code from this perspective. No matter whether it is a feature or a code block, try to determine if it is an optimization or a realization. If it is an optimization, work out whether it is premature or not. Are you working on something that brings little value, or even no value, at the moment? Are you introducing unnecessary complexity? Are you adding more edge cases than actually needed?

The virtue is to avoid complexity. Detect it on time by thinking about premature optimization.

By Igor Spasić

Rules for building systems

This article discusses the application of the Separation of Concerns concept in Build & CI systems. It highlights the difference between a build system and a build management system, the responsibilities of each person and some considerations regarding the consequences when following the rules.

It is structured out as a set of rules to follow when creating build and CI infrastructure for your project.

Think of each rule as a T.A.R.D.I.S.: they are bigger on the inside.

The purpose of this set of rules is to guide us toward C.R.I.S.P. (Complete, Repeatable, Informative, Scheduled, Portable) builds[1].

Distinctions, definitions

To identify the separate concerns, we should provide definitions for the terms:

A **build system** performs transformations in sequence in accordance with a predetermined dependency chain, to create artefacts. A subset of this is the compilation of sources to binaries.

A **build management** system coordinates build system(s).

Specificity

A build system is highly project specific. It is affected by toolchains, project conventions and can generally only run on a specially configured host.

A build management system can run everywhere as long as it can start the build system on the appropriate host. The specificity of a build management system is limited to the number of version control systems it supports. Although it theoretically doesn't have to provide version control support, it is a given that such support will be provided in the minimum feature set.

The Makefile is the build system

Make, rake, Ant, MSBuild, Gradle, grunt etc. are not build systems[2]. You create build systems with them.

Jenkins, TeamCity, TravisCI, BuildBot etc. are build management systems.

The Rules

1. I am Build Server

Rule #1 requires that the build server follows the exact same steps as any other developer.

Expressed the other way around: Every developer has to be able to re-create the complete build process locally, without deviations, when given the development environment and the correct version of the source tree.

2. **When the build server says no, it means no!**

Rule #2 says that if a build server marks a build as broken, then the build is broken. Drop everything and read the logs.

There is no "it works for me", your build server is Judge Dredd: judge, juror and executioner.

You can only adhere to this rule if you have followed Rule #1

3. **IDEs are the enemy A.K.A. F5 is not a build process**

This means that if you drive your development process from an IDE there is no way you can adhere to the Build Rules.

This rule has major consequences regarding the development environment and ties directly into the subject of allowing your developers to use whatever tools they feel comfortable with.

Adhering to the rules

To create a system that adheres to rules #1 & #2, you have to think like a Lego builder: Lots of small, specialized tasks that do one thing and can be used to compose more complex processes.

As an example, doing a *releases* task instead of doing everything in one big implementation will depend on the build tasks for each of the libraries and applications and the tasks that run the tests etc. Using the rake syntax in a contrived example one would do

```
task :release =>[:"test:all"]
task :"test:all" => [:"test:foo", :"test:bar"]
task :"test:foo" => [:"build:foo"]
```

A developer will probably use the component tasks a lot more than the composite release task and we will certainly have a build job on the server that only does releases.

This is a necessity since the system needs to satisfy different usage patterns:

- The build server uses composite tasks that implement complete workflows.
- The developer uses component tasks with surgical precision in the interests of speed and effectiveness.

From the perspective of the build system engineer, this approach is self-evident for the same reason it is evident when building applications: Small chunks of code are easier to manage, test, reuse and understand.

Consequences

The above rules have widespread consequences in structuring the build and CI processes of a project.

The first rule sets the frame within which the build system operates. To adhere to it we avoid using IDE integrations but also build management system integrations (Maven integration in Jenkins being one such example). Handling dependencies, configuring toolchains, and even things like naming conventions are left to the build system. *The build server becomes just another user, performing exactly the same steps a human developer would use.*

The first rule combined with the third lead to the prioritisation of command line usage. This doesn't mean we do everything just from the command line but rather that CLI is the first priority when adding features to the tools comprising the development environment. CLI is the one interface that both humans and bots can operate with the same facility.

The second rule's consequences are a bit more subtle. Avoiding inconsistencies between execution environments is a critical issue and to handle it correctly we need to introduce the concept of a consistent development environment (usually called the 'project VM' as we use virtual machines for encapsulation – although at the time of writing containerisation offers a less resource-intensive approach for specific development scenarios). The challenge of maintaining and replicating such an environment unavoidably leads to the introduction of a provisioning (A.K.A. configuration management) tool such as Chef, Ansible or Puppet.

Who does what

Another way to look at it is that a build system determines the how and what (build, test, package, deploy, release) while a build management system determines the where and when (which CI node, when to trigger etc.).

All of this segues nicely into the final rule:

4. Your (build) infrastructure is a software development project

Rule #4 means you need tests and CI and a plan. You need to budget for CI, for creating a build system specific to your project, and for teaching people how to use it.

To make matters worse, your users are some of the most impatient and downright difficult clients on the face of the planet. They want everything perfect: robust, simple and fast *and* they want it yesterday. You had better be dogfooding by this point.

By Vassilis Rizopoulos

[1] The concept of CRISP builds was first introduced by Mike Clark in his book *Pragmatic Project Automation*
[2] We can debate on CMake and Maven

Successful agile system development with continuous system integration

Continuous integration (CI) is a core practice of successful agile software development. A practice that makes agile systems development, e.g., development of a medical device, successful as well!

Continuous integration of software

A CI infrastructure for software development (hereinafter referred to as „Software CI") enables frequent integration of locally developed source code to the mainline of a software project, even several times a day. At the end of the build process, automated tests are run on different integration levels. Software CI helps in ensuring the consistency of the software and creating a potentially shippable software product on a regular basis.

The same advantages can be achieved by using embedded software as part of a system development project. Embedded software can be developed in an agile way almost without interfering with the development of the mechanical or electronic parts of a device. The electronics of a device – the PCB containing the specific processor – may be under development itself. In this case, the embedded software could first be developed as a simulation running on a PC workstation. In the next step, an evaluation board of the platform vendor could be used as long as the device electronics are not available. In the last step, the software is integrated on the final device's electronics.

Each of these targets (simulation, evaluation board, device electronics) can successively be integrated into the Software CI infrastructure. This ensures that the evolving software runs on the available target.

However, the different engineering disciplines still need to agree on the implementation of functional requirements, non-functional requirements, safety requirements, interfaces, etc. Software CI as part of a system development project is important for ensuring the desired software quality over time. However, CI only on the software level does not really address the risk of failing during (final) system integration – aka "big bang integration".

Continuous integration on the system level

System CI helps with avoiding such "big bang integration" effects. It is far from realistic to achieve the continuity of the fast integration cycles of software, e.g., check-in builds, on the system level. However, the principle of continuously integrating parts to a whole applies at the system level as well. The continuity of software integration is based on the integration of source code parts into the mainline whereas the continuity of system integration is based on the availability of system parts over time, e.g., PCBs, mechanical parts, cables, software functionality, infrastructure, test stands, etc.

The evolution of the System CI is driven by a roughly planned "integration vision". This may start as just a sketch on a whiteboard depicting how the successively available system parts might be put together over time. The integration vision is not a detailed plan, but provides guidance for the integration activities on the system level. The concrete integration of parts is planned and performed as they become available – or maybe a short time in advance. The integration itself is performed as simply as possible. For example, a specially developed sophisticated mechanical fixation is not needed if a tape is sufficient to attach one part to another in an early lab model.

System CI is a constant flow of adding parts and replacing parts with, for example, more mature parts or new revisions of parts. The evolving system is the central target of the System CI infrastructure. The infrastructure takes care of, among other things, building the software, deploying it to the integrated system, and automatically testing the realised system functionality. This allows for regression testing on the system level, which is important in dealing with all the changes on the system integration – as software regression testing is important in dealing with all the changes in software development. The order of particularly meaningful system integration steps needs to be planned, at least roughly. Given a product vision and a draft system architecture, a value-based and risk-based system backlog planning are the basis for successful continuous system integration that lowers the risk of late design breakage and increases the opportunity to develop the right product.

Fail early!

System CI enables us to fail early and thus gives us time to react, change, and test again in fast cycles. Hence, System CI heavily reduces the risk of late design breakage. Additionally, the progress of system development does not have to be deduced or guessed indirectly, but can be demonstrated with the current state of the integrated system – at any point in time along the development lifecycle.

Increase the opportunity to develop the right product by applying this in your next systems project!

By Erik Steiner

The best technology is not always the best choice

As a technology consultant, you are responsible for recommending a technology that is suitable for the customer. That sounds simple, but it isn't. Especially if you get to know about a lot of new technologies at your company, and discuss them with colleagues so you become very familiar with them. The context in which the technology is to be used has to be taken into account. This means that the most advanced technology is not always the right technology for the customer. This also sounds like a trivial statement at first, but when you can see yourself how the latest technology enables the functional requirements X and Z, and fulfils the non-functional requirement Y, the decision can be difficult, and it becomes a real trade-off decision.

Using two real-life examples, I would like to show how difficult this decision can be.

A machine manufacturer wanted to develop new setup and operating software for its highly-specialised machinery. What exactly the data model should look like at the end was not defined, and there was a requirement for an accurate audit of who had made what settings on the machine. We decided to use event sourcing. The primary data source was not a relational database model that mapped the entities, but a list of events that described the change in the system state. This gave us a number of advantages:

- The object model you are working on can be further developed without regard to the past. All you have to do is make sure that the events that are already saved can be applied to the new data model. This means that you can always create it again.
- We guaranteed that we have all required data to provide the audit.

- In contrast to an audit table that is maintained separately, we used the events to intrinsically prevent the developers from forgetting an audit entry.

As already planned from the outset, the machine manufacturer built up a powerful development team of its own, and ultimately took over responsibility for the development. However, the new development team was primarily familiar with the technologies for traditional data-driven applications, and when they took over responsibility for the architecture and further development of the machine software, they were uncomfortable with an architecture they had never seen before. In the end, they switched the application back to a relational database. And they are currently enjoying success with that. In retrospect, we do not feel that we did much wrong, but we apparently underestimated the uncertainty felt by a new team regarding this architecture, and the event sourcing concept was not strictly necessary. It was purely beneficial.

The second example: In 2009/2010, Eric Meijer at Microsoft invented Reactive Extensions. This was initially a .NET library, which made it possible to express events as objects, and complex event sequences as higher-order events. Even early examples demonstrated the broad spectrum of use of this library:

- One example showed how a sensor value becomes a valid alarm via multiple transformations. This means periodically querying values, creating averages using last values, mapping to binary values, and propagating only in the case of changes.
- Another example showed how the requirements for an auto-complete selection box can be implemented. This means that the search query for the suggestions is not submitted if the text is empty or not typed in, and also that a running query is interrupted if further entries are typed in.

- At a very early stage, we also saw the value in gesture recognition, and we were able to identify certain gestures from a sequence of specific movements, and demonstrated this at trade show appearances with the help of Kinect.
- Others have also realised the great value in this library, so it has been ported to numerous programming languages.
- Microsoft has also included the library in the .NET Framework for its Windows phones.

However, the library has enjoyed only limited market penetration over the years, so it was necessary to be mindful of whether the library could be used in all good conscience. The library uses some approaches from functional programming. So, before any event occurs, the entire function chains are defined, in terms of what should happen when an event should occur. These events are provided as observables, and the events are only generated or observed when a consumer of these observables is interested in the potential outcomes. As powerful as this concept is, it can cause headaches for an untrained developer.

There was a case with one of our projects, in which the usage of Reactive Extensions was removed because the developer, who had to fix a complicated bug, did not understand the code. In the end, the bug was somewhere else, but the code kept removed.

There is a positive example of one customer using Reactive Extensions as an internal API for web services. That was risky, because although the solution was technically superior to the alternatives, it was not a common approach at that time. It was only justifiable because the customer already had a stable development team that could support the decision.

The example also shows that a good solution takes years and marketing to become mainstream. Google has used the JavaScript variant of Reactive Extensions (RxJS) in the Angular2 framework to do just that – but 5 years later. There are now lectures at numerous conferences about RxJS and how it can be used. Google has helped to bring it out of its niche. You can now recommend this technology even if there will not be long-term access to the development team to provide training in it.

So, if a technology or architecture is interesting because it is suitable for the problem but is niche or still upcoming, you need to make sure that the development team understands the technology, recognises its benefits and believes in it. Because the team will not get much confirmation that it is using the right technology.

By Carsten Kind

Watch your state

In May 1960, the legendary Peugeot 404 was presented to the public. Its manufacturing process was controlled by manually curated lists in which each car in the production pipeline had its particular entry. Each car was assembled according to a fixed plan. Consequently, spontaneous adaptations or dynamic modifications of the production plan were impossible when components were suddenly missing, or when experts responsible for certain tasks got sick.

It's all about state

The main factor enabling the transition from the *centralised* setup of the early 60s to today's *decentralised* Industry 4.0 is the way in which state is handled: today, each car body automatically knows its exact specification and configuration. Such local state provides a great deal of flexibility: the order in which cars are assembled can be adapted at any time. No central list or global database needs to be queried or updated. Fewer central points of failure and bottlenecks exist. Moreover, cars and other manufactured „smart" items can be connected, they can identify each other, and they can exchange information using a decentralised ad hoc mesh-up topology. The exchange between the manufactured items is even possible during the production process. Thanks to early feedback, fewer dependencies, and more independent production units, it is possible to achieve early adaptation and just-in-time production. Consequently, production quality is significantly improved and costs are reduced.

State in today's communication protocols

Turning from industrial production to Internet communication protocols, the goals of improving quality and reducing costs translate to reducing latency and lowering energy consumption:

1. When Internet user Alice connects to her online bank, she needs to authenticate herself before accessing her savings and transactions. To initiate such an authenticated session, she provides unique credentials (login and password, and hopefully a second authentication factor). If the login is successful, the bank grants Alice an access token valid for that very session. Alice can then seamlessly re-authenticate herself during the session using that token.

   Such a token contains information about the session, in particular, information to identify the user associated with the session. The token thus represents session *state*. The question we ask is where and how the bank should manage such temporary access tokens? The bank could store a copy of millions of such session tokens and perform equality checks with the tokens that Alice sends with her requests. If Alice's token exists in the list of copied tokens, the bank can safely assume it's indeed Alice who is connecting.

This, however, is very inefficient as the bank's server needs to perform several memory lookups for each customer request to fetch the corresponding token. Unfortunately, most web application firewalls (WAFs) today work exactly this way.

Before discussing smarter ways of handling access tokens, we will look at two more examples.

2. Packet forwarding in today's Internet: when Alice sends a data packet to Bob, she adds Bob's IP address to the packet header and hands the packet over to her Internet provider. From then on, every router on the path from Alice to Bob extracts Bob's address from the packet to do a database lookup to retrieve information about the next hop, that is, where to forward the packet to. Each router bases its forwarding decision on local *state* that was previously learned from neighbouring routers.

   This procedure is not only insecure (as the neighbours' information is not properly authenticated), but also very energy-consuming and inefficient (as millions of packets per second need to be looked up). Good news: a more secure and – at the same time – more efficient solution exists.

3. Finally, TCP's vulnerability against denial-of-service attacks: in the initial step of the three-way handshake between client and server, the server needs to store information about the client in order to remember the client when it completes the handshake. The storage of even small amounts of *state* can lead to attacks, when huge numbers of clients pretend to connect to the server, but execute only the first step of the three-way handshake. The server cannot

know upfront if a client is malicious and thus keeps waiting for each client to come back to complete the handshake. Soon, when many clients connect simultaneously, the server's resources are exhausted and new connections cannot be accepted. Users then perceive the server as unavailable, aka under a denial-of-service attack.

Higher latency, reduced throughput, increased energy consumption, DoS attacks – what is the solution to prevent these design flaws in authentication, packet forwarding, and TCP? All examples have in common a non-optimal handling of state that is frequently consulted as part of mission-critical business processes.

Remediation

A crucial insight for remediation is that state should be stored by the service *requester*, not by the service *provider*. Examples include waiting stamps at the post office counter: they are stored by the waiting customer, not by the post department. Except for a single global counter, the post department keeps no state for each customer waiting in line.

A very similar solution works for web authentication: The server should not remember session state for each client, but send to the client a cryptographically signed token with information about the session state. The client presents such token whenever it connects to the server, who then verifies the token's validity. Instruction sets on modern CPUs verify authenticated AES messages within roughly 50 cycles (20ns at 2.5GHz). For comparison: a DRAM memory lookup takes up to 200 cycles, and a round trip in a data centre takes up to 500,000ns.

Packet forwarding should ideally work similarly: each packet should contain cryptographically protected information about its path to the destination. Each router on the path checks such forwarding information by efficiently verifying an AES-MAC, instead of wasting energy and time on costly database lookups. The efficacy of this solution is demonstrated by the Internet architecture SCION.

Finally, TCP's vulnerability can be fixed by so-called SYN cookies, i.e. sequence numbers that encode information about the client so that state at the server can be significantly reduced. Due to space constraints, we refer to the literature for more information.

To conclude, the storage locations of all application state are crucial for enabling efficient system processes. Ideally, state is treated as a first-class citizen and outsourced to users and service requesters. In order to avoid abuse of functionality based on the outsourced state, efficient cryptography is required. Fortunately, today's cryptography not only makes the entire application more secure, but – perhaps surprisingly at first glance – also more efficient.

By Raphael M. Reischuk

You always have time for a proper root cause analysis

A production issue will pop up, even in solid software. Don't panic.

We, as software engineers, are more than eager to resolve the issue quickly. We should be, yet we should never stop being careful and analytical. Otherwise we might deploy another embarrassing bug or suppress the problem without solving it, making everything worse.

Investing an hour in analysis is always worth it.

People will describe the symptoms

Keep in mind that users will report *symptoms*. This has nothing to do with a lack of understanding or skill. They have an outside view.

What you need to find, and then fix, is the *problem* (or, more precisely, its *cause*).

Don't jump to conclusions. Listen to all the clues. Analyse. Reproduce. Rethink. Refine (or create) the ticket. Remember that every bug is a strong indicator of a missing test, so think of test cases which are explicit about expected and actual results. And write tests to cover your findings.

Don't have someone else do the analysis for you; you need to be part of the thought process. This might very well take longer than the coding. But if you don't do this slowly and deliberately, you might fix only half the problem or even make it worse. Don't be that cowboy.

Can we be more specific?

Absolutely!
Let's have a look at some of the painfully obvious yet all too common pitfalls.

Not really fixing the issue. Or: Not fixing the real issue.

Do you feel some guilt reading through the three examples below? I do.

- Added a fix in the wrong place. Didn't solve the problem, optionally introduced another one?
- Fixed a NullReferenceException by adding another if, ended up with no exception but a wrong result?
- Divided an x through total to make it relative, deployed to production, got an „unexpected" DivideByZeroException the day after?

Things are often more subtle and not that silly. What the three examples have in common though is that we didn't think it through all the way. We knew what *shouldn't* happen, but didn't find out *why* it did and what *should* happen.

"We cannot reproduce this!" or "It's expected behaviour."

It's a chance to improve the error message.

You couldn't find customer #316 in the DB, because it actually doesn't exist? A NullReferenceException with a stack trace is still a bug. Tell them what's wrong, in their language, so everyone can understand the problem without a debugger.

Also do this when you cannot reproduce a problem. Don't try to solve it based on assumptions; refine your error handling and use it to find the problem.

... This might have happened somewhere else

„I think we might divide by total there, there and there too. We should check if that also throws a DivideByZeroException."

It's great to think about it. The problem might be a pattern. But fix that separately. Equally carefully!

Does this happen too often? That's what the DRY principle is about.

There's a "real solution" and a "quick solution" (and a "workaround")

After understanding the problem, there might be multiple options to solve or work around it:

1. adding another if in method x
2. refactoring the accounting system interface to support zero totals
3. breaking down the system into different microservices

It's fine to add an if to get an imminent problem out of the way – after you have thought it through! In any case, keep all the options in your backlog. After adding the 5th if, another solution could be increasingly compelling.

Conclusion

All issues, even seemingly trivial ones, demand careful and thorough analysis. The problem might seem obvious, but it's always worth double-checking. You always have time for that.

Having said that, be pragmatic with the fix. Don't over-engineer. Decide what's needed *now* to put out the fire, and what should be done *later* to prevent further blazes.

But let it be an educated decision. And always test thoroughly.

By Matthias Meid